THE
MONO BASIN
ECOSYSTEM

Effects of Changing Lake Level

Mono Basin Ecosystem Study Committee
Board on Environmental Studies and Toxicology
Commission on Physical Sciences, Mathematics,
and Resources
National Research Council

National Academy Press
Washington, D.C. 1987

NATIONAL ACADEMY PRESS • **2101 Constitution Avenue, NW** • **Washington, DC 20418**

NOTICE: The project that is the subject of this report was approved by the Governing Board of the National Research Council, whose members are drawn from the councils of the National Academy of Sciences, the National Academy of Engineering, and the Institute of Medicine. The members of the committee responsible for the report were chosen for their special competences and with regard for appropriate balance.

This report has been reviewed by a group other than the authors according to procedures approved by a Report Review Committee consisting of members of the National Academy of Sciences, the National Academy of Engineering, and the Institute of Medicine.

The National Academy of Sciences· is a private, nonprofit, self-perpetuating society of distinguished scholars engaged in scientific and engineering research, dedicated to the furtherance of science and technology and to their use for the general welfare. Upon the authority of the charter granted to it by the Congress in 1863, the Academy has a mandate that requires it to advise the federal government on scientific and technical matters. Dr. Frank Press is president of the National Academy of Sciences.

The National Academy of Engineering was established in 1964, under the charter of the National Academy of Sciences, as a parallel organization of outstanding engineers. It is autonomous in its administration and in the selection of its members, sharing with the National Academy of Sciences the responsibility for advising the federal government. The National Academy of Engineering also sponsors engineering programs aimed at meeting national needs, encourages education and research, and recognizes the superior achievements of engineers. Dr. Robert M. White is president of the National Academy of Engineering.

The Institute of Medicine was established in 1970 by the National Academy of Sciences to secure the services of eminent members of appropriate professions in the examination of policy matters pertaining to the health of the public. The Institute acts under the responsibility given to the National Academy of Sciences by its congressional charter to be an adviser to the federal government and, upon its own initiative, to identify issues of medical care, research, and education. Dr. Samuel O. Thier is president of the Institute of Medicine.

The National Research Council was organized by the National Academy of Sciences in 1916 to associate the broad community of science and technology with the Academy's purposes of furthering knowledge and of advising the federal government. Functioning in accordance with general policies determined by the Academy, the Council has become the principal operating agency of both the National Academy of Sciences and the National Academy of Engineering in providing services to the government, the public, and the scientific and engineering communities. The Council is administered jointly by both Academies and the Institute of Medicine. Dr. Frank Press and Dr. Robert M. White are chairman and vice chairman, respectively, of the National Research Council.

Support for this project was provided by contract 05-85-01 between NAS and the U.S. Forest Service, an agency of the U.S. Department of Agriculture.

Library of Congress Catalog Card Number 87-62016
ISBN 0-309-03777-8

Printed in the United States of America

Cover Photograph by David Policansky

Preface

The Mono Basin of California, with Mono Lake at its center, is an area of unique aesthetic appeal and scientific interest. As such, it has been designated as a national scenic area. It is also an important water resource, with basin water being diverted for use in Los Angeles. If water levels in Mono Lake were to change--whether as a result of this diversion, natural phenomena, or a combination of these factors--a variety of complex changes in the Mono Basin ecosystem could occur. Concerns about these effects led to a directive by Congress for review of the pertinent scientific information. The National Research Council formed the Mono Basin Ecosystem Study Committee to carry out this task, and funding for the study was provided by the U.S. Forest Service.

The congressional directive, found in Public Law 98-425 and House Report 98-291, notes that the study is intended to include, but not be limited to:

(1) an inventory of all terrestrial and aquatic species, indicating their population dynamics, historic and current population levels, and probable trends as to future numbers and welfare;

(2) the critical water level of Mono Lake needed to support current wildlife populations;

(3) the hydrology of Mono Lake, including ground water inflow, evaporation and fresh water spring inflow,

and water balance at the critical water level, showing the estimated evaporation and projected inflows;

(4) the estimated wildlife populations using Mono Lake which would be supported at the estimated water levels that would occur as the City of Los Angeles continues to exercise its water rights as such rights have been granted or may be modified under the laws of the State of California;

(5) the significance of any changes from current wildlife populations to those as may be estimated based upon such study and referenced to the populations of such wildlife in other areas.

In addition, the U.S. Forest Service requested that the committee consider issues related to the management of the scenic area, including the effects of fire and grazing on the ecosystem, effects of increased public access on the tufa formations, effects of lake levels on air quality, and an inventory of vegetation types in the basin. The committee was specifically asked not to address the socioeconomic issues of water rights or the details of the management of the water.

In carrying out its task, the committee consulted with state, federal, and local agencies and with scientists conducting research in the Mono Basin. A small number of scientists have studied the Mono Basin, and the committee relied heavily on their work. Much of the available information appears in the form of draft manuscripts and unpublished reports that have not been subjected to peer review. The committee has independently evaluated much of this literature and cautiously accepted many of the data reported in these unpublished documents. In some cases, crucial information was not available in any form. The limitations of the data are discussed throughout the report.

Because of these limitations and because of the committee's awareness that many people consider the Mono Basin area to be almost sacred while others consider it to be a source of exploitable resources, the committee approached its task with caution but an enthusiastic willingness to do the job properly. This report is the product of many days of doing research, exchanging ideas, deliberating, and writing.

Mono Lake and the surrounding national scenic area managed by the U.S. Forest Service have become the subject of a widespread controversy. Throughout the country, but especially in the West, there are bumper stickers advocating "Save Mono Lake." Conservation groups continue to use legal pathways in their attempts to preserve scenic and ecological components of the basin in the face of perceived threats to its resources. This controversy fueled the need to know more about the integration of the hydrological, physical, chemical, and biological components of the basin. If we, the public, are going to continue to make demands on the resources of the Mono Basin, we need to know the risks we are taking with the ecological relationships in the basin ecosystem that have been established over centuries, if not millennia. This report on the Mono Basin is thus an ecological risk assessment study of a complex natural ecosystem.

The aesthetic appeal and scientific interest of the Mono Basin derive from its unique collection of natural wonders. Few places record the history of Pleistocene glacial advances and retreats more clearly than do the numerous and massive moraines of the Mono Basin. The basin also contains an array of readily observable volcanic features that are as dynamic as any on earth. Here occur deep stacks of basaltic flows, massive rhyolitic domes, recent cinder cones, pumice blocks that float, and numerous and widespread ashfalls spanning over 700,000 years in age.

The basin straddles the transition between two very dissimilar physiographic provinces, the Sierra Nevada and the Great Basin Desert. Elevations range from 6,380 ft (the current elevation of Mono Lake) to over 13,000 ft. The changes in climate, soils, and biota over this elevation range are equivalent to those found in traveling over thousands of miles of latitude.

Mono Lake lies immediately east of the high Sierra. The lake thus receives very little rainfall or snowfall, while the high elevations of the Sierra collect large amounts of snow, which melts to supply water to the lower basin.

The basin has cultural as well as scientific interest. The indigenous Owens Valley Piutes utilized the natural resources of the region. They were a sophisticated seed-collecting society.   Their seed-milling sites can still be found atop granite boulders along the valley and in the foothills.

The beauty of the Mono Basin is impressive, and the members of the committee were not untouched by its aesthetic appeal.   They realized, however, that this report could not reflect any personal feelings about the lake or basin but must present the most unbiased scientific analysis possible.   For a short period during our final meeting, I asked committee members to discuss "what the lake means to them."   Their comments demonstrate that they are as observant about aesthetic features as they are about scientific features.   I share the following representative comments with the reader as somewhat of a counterpoint to the scientific discussion in the report itself: the lake setting, with the natural wonders of both desert and mountains close at hand, is particularly impressive; the lake has a magical quality--it is not just another big, blue lake; the lake in its present configuration appears to be balanced with its setting; the tufa towers with their reflection in the lake are aesthetically more pleasing than tufa towers on land--the lake-tufa tower relationship characterizes Mono Lake; the interplay of tufa towers and birds represents the character of the lake; and some amount of playa (exposed lake bed) gives a definition to the lake and is pleasing; the larger islands, as discrete islands, are part of the aesthetic balance of the lake; the current numbers and diversity of birds symbolize the biological uniqueness of the basin.

This report can be used for many purposes.   It can be a guide to resource managers or a demonstration to naturalists and ecology classes of the integration of complex interacting factors in a defined but broad ecosystem.   As a case study in ecological risk assessment, it might be used in college classes and seminars or as an example of a systematic assessment of the effects of water diversions on similar systems.

The report stresses the need to continue monitoring the ecosystem to test predictions as the basin continues to be perturbed.   If critical conditions are approached that may trigger the catastrophic ecosystem responses identified in the report, studies should be initiated to test the accuracy of the predictions.

This report, although it has its limitations, contains a wealth of information about the Mono Basin ecosystem. The committee hopes it will be a major contribution to our understanding of the basin and perhaps other hypersaline, closed-basin lakes. I hope that it proves useful to those concerned about the future of Mono Lake and the Mono Basin.

Duncan T. Patten
Chairman,
Mono Basin Ecosystem Study
Committee

Acknowledgments

There are many organizations and individuals the Mono Basin Ecosystem Study Committee wishes to acknowledge. John Roupp, Nancy Upham, Dick Warren, Carl Westrate (now retired), and others with the U.S. Forest Service have continuously given us information and support. Mel Blevins, Chris Foley, Eldon Horst, LeVal Lund, Andy Pollack, and other personnel from the Los Angeles Department of Water and Power have been cooperative throughout this study. They have shared LADWP data with committee members to facilitate our understanding of ecosystem processes at Mono Lake. Martha Davis, David Gaines, and others from the Mono Lake Committee were also helpful throughout the study. Dan Botkin and the California Fish and Game Mono Lake Blue Ribbon Panel expanded our information base through sharing of data. Individual researchers that have helped the committee through interpretation or sharing of information include: Scott Stine (paleogeomorphology); Peter Vorster (hydrology); Joe Jehl and David Winkler (ornithology); Gayle Dana, David Herbst, and Robert Jellison (aquatic biology); Kerry Sieh (geology); and Ron Oremland and Ronald Spencer (geochemistry). Many individuals reviewed the report, and to them we are grateful for their suggestions on how it could be improved.

The Mono Basin Ecosystem Study Committee is thankful for the herculean efforts of so many of the staff members of the Board on Environmental Studies and Toxicology (BEST). Ruth DeFries, staff officer, has supported, prodded, and cajoled us in order to produce this report.

Her patience and professionalism have been appreciated by all committee members. David Policansky, staff officer, has been our conscience, keeping us honest and on the point. Without administrative secretary Tracy Brandt, all deadlines would have fallen by the wayside. She has brought together many loose ends and helped fill the staffing gaps whenever called upon. Roseanne Price, of the editorial staff of the Commission on Physical Sciences, Mathematics, and Resources, handled editing of the drafts and assisted with production. Thanks also go to Myron Uman, previous director of the Environmental Studies Board, who introduced us to our task, and to Devra Davis, staff director of the Board on Environmental Studies and Toxicology, who through her interest gave us encouragement at some difficult times.

The committee is indebted to all of these people and organizations, and to the U.S. Forest Service for providing the financial support and the National Research Council for providing the opportunity to participate in this study.

Contents

Executive Summary

The Mono Basin National Forest Scenic Area lies in eastern California about 300 mi north of Los Angeles and 190 mi east of San Francisco. The basin is walled in by the eastern escarpment of the Sierra Nevada to the west and by Great Basin ranges to the north, south, and east. Consequently, no water naturally flows out of the basin. The only loss of water occurs through evaporation and, since 1941, diversions of fresh water by the city of Los Angeles.

The Mono Basin is the hydrologic drainage basin for Mono Lake. As a result of millennia of evaporation from its surface, the lake (at 500,000 years one of the oldest in North America) has gradually increased in salinity, and it is now about 2.5 times as saline as the Pacific Ocean. The salinity--and the particular chemistry of the dissolved ions--has a large effect on the biota of the lake and the ecology of the basin. The unusual chemistry of the lake is also responsible for one of its distinctive scenic attractions, the tufa towers.

Mono Lake has a simple but productive ecosystem. Benthic and planktonic algae provide the major base of the food chain in the lake. The primary consumers are aquatic arthropods, mainly the pelagic brine shrimp (*Artemia monica*) and the benthic brine fly (*Ephydra hians*), which are present in immense numbers. Both brine flies and brine shrimp depend on algae for their food. In addition, the brine fly requires shallow habitats for feeding, and shallow, hard surfaces for reproduction.

1

The great productivity of these aquatic invertebrates permits hundreds of thousands of birds to use Mono Lake. In addition, no fish or other aquatic predators live in the lake to compete with the birds for the abundant food supply. The most abundant birds on the lake (with estimates of maximal population numbers using the lake annually) are the eared grebe, *Podiceps nigricollis* (900,000); Wilson's phalarope, *Phalaropus tricolor* (125,000); the red-necked phalarope, *Phalaropus lobatus* (54,000); and the California gull, *Larus californicus* (50,000). The populations represent one-quarter to one-third of the North American population of eared grebes and 15 to 25 percent of the North American population of California gulls. Eared grebes and red-necked phalaropes use the lake as a stopover during migration; Wilson's phalaropes use it as a major staging area before undertaking a long, possibly intercontinental, flight; and California gulls nest on the islands in the lake.

The Los Angeles Department of Water and Power (LADWP) has been diverting fresh water from streams that feed Mono Lake since 1941. As a result, the lake level has dropped about 40 ft to approximately 6380 ft above sea level in 1986. Because of the potential ecological effects of water diversion, there has been much concern over the future of the lake. In 1984, the Congress passed legislation designating the area as a National Forest Scenic Area, the only such area in the country. The same legislation mandated this study by the National Research Council to review the available scientific information and to assess the potential effects of changing lake levels on the ecosystem of the Mono Basin.

Various resources of the Mono Basin ecosystem--aquatic biology, bird populations, and shoreline and upland environments--are here assessed (see figure on the following page). The scientific background for these assessments is presented in chapters 2 through 5 of the report, and the effects of changes in lake level are discussed in detail in chapter 6.

Because the various resources would be affected differently by different lake levels, and because the effects occur gradually over a range of levels rather than at one specific level, the consequences of changing lake level are

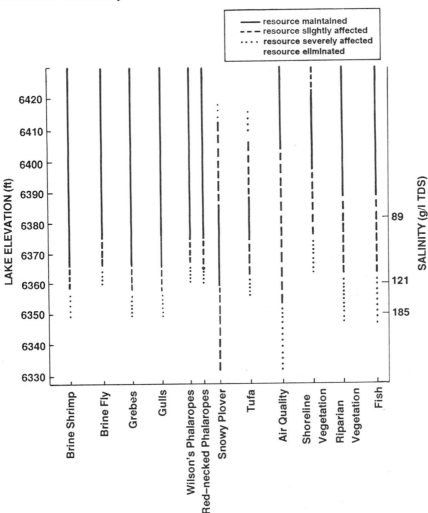

assessed for a series of lake levels. Lake levels from the elevations of 6430 to 6330 ft above sea level, in intervals of 10 ft, are considered. The upper level of 6430 ft was chosen because it is close to the historic high stand of 6428 ft reached in 1919. This level could conceivably be reached again after a series of wet years if no water were exported from the basin. The lower level of 6330 ft was chosen because it is the approximate stabilization level (i.e., inflows of water would equal loss through evaporation), assuming exports of 100,000 acre-ft/yr of water from

the basin and climatic conditions similar to those of the
past 40 years. Most of the ecological consequences of
water levels below 6330 ft would presumably not be significantly different from those of a water level of 6330 ft.

Because there are no natural outlets from Mono Lake,
ions dissolved in the lake would become diluted, i.e., salinity would be reduced, if water level rose and increased
the volume of water in the lake. Salinity would increase if
lake level fell. Approximate salinities for lake levels between 6330 and 6430 ft are given in Table 6.2. The salinity of lake water critically affects the ability of aquatic
organisms to thrive, and therefore salinity is a crucial factor in the lake's ecosystem.

The depth and salinity of Mono Lake encourage stratification of its waters, which in turn affects chemical and
biotic processes. The wet winters of 1983 through 1986
have resulted in large influxes of fresh water, which has
only partially mixed into the heavier saline lake water.
The result has been meromixis, an incomplete mixing of the
lake's waters. Meromixis could have profound effects on
the chemistry and the biology of the lake by trapping nutrients in the bottom layer and could be expected to inten-
sify if inputs of fresh water are large in relation to the
volume of saline water. If salinity increased above approximately 125 g/l of total dissolved solids (TDS), minerals
that contain sodium would begin to precipitate and a persistent stratified layer would be likely to occur near the
bottom of the lake. The precise effects of meromixis and
precipitation of minerals on the salinity cannot be quantified with the current understanding of the geochemistry
of Mono Lake. Meromixis, with a less saline surface layer,
could to some extent alleviate the effects of increasing
salinity for the lake's biota.

The algae in Mono Lake are fairly resistant to increased
salinity, although their productivity is likely to decrease
gradually at salinities above about 100 g/l and decrease
more rapidly above 150 g/l (corresponding to lake levels of
approximately 6370 ft and 6350 ft above sea level). Brine
shrimp are expected to gradually decrease in abundance if
salinity exceeds 120 g/l (corresponding to a lake level of
approximately 6360 ft) because of bioenergetic demands
placed on larval growth and development and reductions in

primary productivity. The effects on brine shrimp populations would be severe if salinity reached 150 g/l (lake level of approximately 6350 ft). Reduction in brine fly populations would be large at salinities greater than 130 g/l (lake level of approximately 6356 ft).

Changes in lake level would also affect the habitat available for brine flies to reproduce and feed. If the lake level declined to approximately 6370 ft, the acreage of shallow, hard substrate would be reduced from current values by about 40 percent, leaving considerably less habitat area. The brine fly population would consequently be affected, although it is not possible to predict the precise impact.

Although increased salinity does not appear to physiologically affect birds that use the lake, a decreased food supply would certainly limit their numbers. The critical food resources for aquatic birds using Mono Lake are brine shrimp and brine flies. If the lake fell to levels at which the birds' food sources were adversely affected, the bird populations would be reduced. The decrease in availability of brine shrimp for food would begin to affect those birds relying on them--eared grebes and California gulls--at a salinity of 120 g/l (lake level of 6360 ft). The impacts would be acute at salinities above 150 g/l (6350 ft). For those birds relying on brine flies--the phalaropes--impacts would begin at a lake level of 6370 ft and would become acute at levels below 6360 ft.

Lower lake levels would also reduce the surface area of islands that are free from predators and hence available to California gulls for nesting sites. At a lake level of approximately 6350 ft, virtually all islands would be connected to the shore, and those gulls remaining despite the loss of lake-supported food would not be able to nest at the lake.

Changes in lake level would also influence the shoreline environment, notably the vegetation, snowy plover habitat, tufa formations, and air quality. In general, changes in shoreline vegetation would be controlled by changes in the availability of fresh water and inundation of suitable habitat as lake level fell or rose. A rise above the present level would inundate shoreline vegetation. A drop in lake level would expose additional barren areas of playa, and vegetation would be established only where springs and seeps provide fresh water. If lake level fell below the current level, streams would incise, reducing or eliminating established vegetation and denuding their banks.

The snowy plover (*Charadrius alexandrius*), the only shorebird that would be affected by changes in lake level, numbers about 350 birds at Mono Lake, about 11 percent of the California population. The plovers nest primarily on the exposed playa and pumice dunes on the lake's eastern shore and have probably benefited from recent drops in lake level because additional areas of playa have become available for nesting. If lake level rose, the playa area would be reduced. However, even complete inundation would probably not destroy all their nesting, because there is nesting area available above the playa. Lower lake levels would expand the availability of habitat; a lake level of approximately 6360 ft would probably maximize the nesting population.

A drop in lake level would expose currently submerged tufa towers and permit public access to towers that are now offshore. There would be the potential for increased visibility of the tufa for visitors; there would also be the potential for increased vandalism. A rise in lake level would reduce access to tufa towers. Wave action could damage some of the tufa towers and would start to destroy some of the delicate sand tufas if lake level rose above 6390 ft.

Falling lake levels would expose more of the lake bed and would increase the supply of alkaline dust to form dust storms during high winds, thus increasing the frequency and severity of dust storms. Conversely, a rise in lake level would decrease the frequency and severity of dust storms.

Most features of the upland environment of the Mono Basin would not be affected by changes in lake level. However, riparian (streamside) plants and animals, although not directly affected by changes in lake level, would be affected by changes in streamflows associated with changes in lake level. Minimal flows of 19 cubic feet per second (cfs) and 10 cfs are currently being maintained by court order in lower Rush Creek and lower Lee Vining Creek, respectively. The two hydrologic models of the basin predict different lake levels, 6360 and 6330 ft, that would be maintained by these flows. These flows should generally be adequate to maintain riparian stands of vegetation equivalent to those present in 1941, before diversions began. On the lower part of the alluvial fan, however, downcutting has destroyed the riparian community where recruitment might otherwise have occurred. These flows are also adequate to maintain breeding populations of brown trout. Thus, all flows necessary to maintain the lake level above approximately 6360 ft, regardless of which model is used, should maintain riparian vegetation and fish populations in and alongside lower Rush and Lee Vining creeks. Periodic heavy releases in spring are necessary to enhance the recruitment of riparian vegetation.

Responses of various resources to changes in lake level will, for the most part, occur gradually over a range of levels. (These consequences of changes in lake level are summarized in Table 6.13, and the range of levels over which they are predicted to occur is summarized, with three salinities added for reference, in Figure 6.3, which is reproduced on page 3 of this executive summary.) Decisions about optimal lake levels to protect the ecosystem will require decisions about which resources or combinations of resources are most important. Undoubtedly, trade-offs between preservation of different resources will have to be made.

Climatic fluctuations and long-term changes will produce fluctuations in lake levels. Because the predictions in this report are based on data and understanding that are incomplete or uncertain in some cases, the committee cannot pinpoint precise lake levels at which particular effects will occur. For all these reasons, if a maintenance lake level is selected, the committee strongly recommends that a buffer be built into that level to protect against such uncertainties.

1
Introduction

Mono Lake lies at the heart of the Mono Basin in eastern California. The basin supports a variety of wildlife, some of which is dependent on the lake ecosystem. Since 1941, the city of Los Angeles, which owns the water rights to several of the major streams that flow into the lake, has been exporting water from the basin. Between 1941 and 1985, exports averaged 68,100 acre-ft/yr. Between 1969 and 1985, exports generally increased and averaged 90,100 acre-ft/yr (Los Angeles Department of Water and Power (LADWP), 1987). As a consequence of these water diversions, the lake level has dropped about 40 ft since 1941. Questions have been raised about the effects of changes in the water level of Mono Lake on the basin ecosystem.

With the passage of the California Wilderness Act in 1984, Congress established the Mono Basin National Forest Scenic Area and placed management of the basin under the jurisdiction of the U.S. Forest Service. The Bureau of Land Management previously had had responsibility for managing the area. Since 1982, the state of California has managed the land exposed since 1941 by declining lake levels (land below an elevation of 6417 ft above sea level), designated the Mono Lake Tufa State Reserve.

The California Wilderness Act of 1984 also contained a request for the National Research Council to review the available scientific information and assess the current understanding of the effects of changes in lake level on the scenic area's ecosystem. This report presents the results

8

of that study, which was carried out by the Mono Basin
Ecosystem Study Committee of the NRC's Board on Environmental Studies and Toxicology.

The hydrologic drainage basin from which water flows
into Mono Lake includes all of the land area of the Mono
Basin National Forest Scenic Area (Figures 1.1 and 1.2).
Changes in lake level, which in recent times have resulted
primarily from diversions of most of the streams carrying
freshwater runoff from the eastern escarpment of the Sierra Nevada to Mono Lake, affect the lake system itself, the
shoreline and upland portions of the basin, and wildlife in
the basin.   No streams drain the lake, so the amount of
water in it is determined by inputs from rainfall, snowmelt,
and springs and loss from evaporation.   When inputs are
reduced, either by changes in the climate or diversions, the
lake level drops.   The ions dissolved in the water are concentrated as water evaporates from the lake surface, leading to increases in salinity with lower lake levels.   At
higher levels of salinity, the reproduction and survival of
the lake's biota, mainly brine shrimp and brine fly, would
be affected and the birds that rely on the brine shrimp
and brine fly for food would in turn be affected.   Other
concerns about physical changes in the lake have been
raised.   In particular, some of the lake's islands would become peninsulas if the lake level were to fall.   Predators
could then prey on birds that customarily nest on the
islands.

Changes in lake level would also have consequences for
other parts of the basin.   Along the shoreline, airborne
dust would increase if lake level fell and exposed alkaline
flats.   Rising lake levels could result in damage to the tufa
towers, the pillarlike formations that are a distinctive scenic resource of the basin.   Changes in lake level might
also alter the depth of the water table and consequently
the shoreline vegetation.   In the upland portions of the
basin, altered streamflows associated with changes in lake
level would affect the riparian vegetation and the fish and
other wildlife relying on that habitat.

The congressional directive, House Report No. 98-291
(U.S. Congress, House of Representatives, 1983), specified
that the NRC study on the effects of changing lake levels

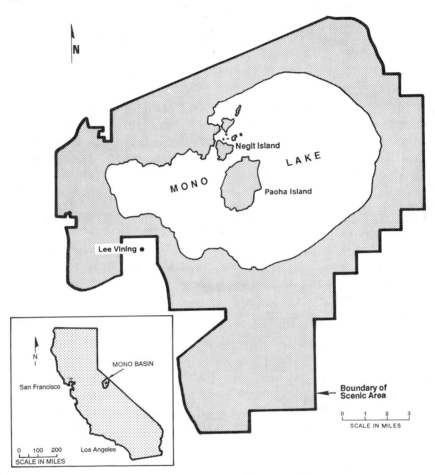

FIGURE 1.1   Mono Basin National Forest Scenic Area (courtesy of the U.S. Forest Service).

on the ecosystem include, but not be limited to, the following:

(1)   an inventory of all terrestrial and aquatic species, including current and probable future population levels;
(2)   the critical lake level needed to support current wildlife populations;
(3)   the hydrology of Mono Lake;

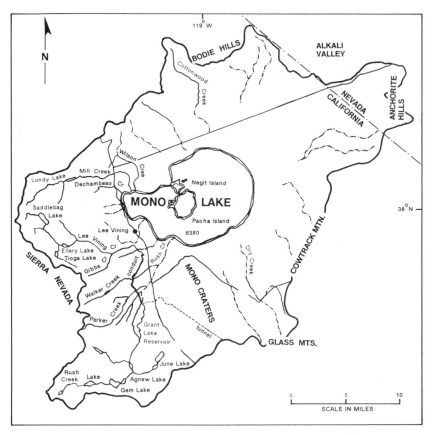

FIGURE 1.2   Hydrologic drainage of Mono Basin (from Vorster, 1985).

(4)   the estimated wildlife populations if Los Angeles continues to exercise its water rights; and

(5)   the significance of changes in wildlife populations to populations in other areas.

In addition, the legislation specified that this study should not address the socioeconomic issue of Los Angeles's water rights.   The U.S. Forest Service requested that the committee also consider issues related to the management of the scenic area, including the effects of fire and grazing on the ecosystem, effects of public access to tufa formations, effects of lake level on air quality, and an inventory of vegetation types in the basin.

Because different lake levels have different consequences for the various resources of the Mono Basin (e.g., lake biota, bird populations, tufa towers, shoreline vegetation, and riparian habitat), the committee approached its task by analyzing a series of lake levels both above and below the current level. Chapter 6 of this report discusses in detail the consequences of each of these lake levels. Chapters 2 through 5 discuss the available literature and provide background information for the analysis in chapter 6. Chapter 2 describes the hydrology of the Mono Basin; chapter 3, the physical and chemical lake system; chapter 4, the biological system of the lake; and chapter 5, the shoreline and upland systems. In the remainder of this introduction, the climatology, physiography, and geology of the Mono Basin and the prehistoric and historic fluctuations in lake level are briefly described, and Mono Lake is compared with other saline lakes.

CLIMATOLOGY, PHYSIOGRAPHY, AND GEOLOGY OF THE MONO BASIN

Mono Basin, which lies in eastern California approximately 190 mi[*] east of San Francisco and 300 mi north of Los Angeles, is a closed hydrologic basin walled in by the steep-faced eastern escarpment of the Sierra Nevada on its western side and by Great Basin ranges on the north, south, and eastern sides (Figure 1.2). Within the basin is

---

[*]The reader will note that no attempt has been made to impose consistency of style on units of measure throughout the report. The use of English units of measure is still conventional in many of the disciplines central to the subject of this report, but when the original measurements were made in metric units they were retained in that form.

Mono Lake, known as a terminal lake because under natural conditions all the runoff and groundwater seepage from the basin terminate in the lake.

Estimates of the areal extent of the basin range from 634 to 801 mi^2. The differences are attributable to the difficulties of interpreting the drainage divide in the eastern and southern parts of the basin and the inclusion of 69 mi^2 of alkali flats in Nevada that might have subsurface connections to the basin (Vorster, 1985).

The Mono Basin lies on the border of two major physiographic provinces--the Sierra Nevada and the Great Basin. The topography of the basin varies greatly. Elevations range from approximately 6,380 ft above sea level at the lake surface (in August 1986) to approximately 13,000 ft at the crest of the Sierra Nevada. The basin is characterized by large seasonal and annual variability in precipitation. A majority of the precipitation occurs in winter in the form of snow. Both amount of precipitation and temperature vary considerably in the basin as a function of elevation and distance from the Sierra Nevada. The variability in precipitation is attributed in part to the complex effect of the mountains on Pacific storms. Episodes of strong winds occur throughout the year but are most frequent in the late fall through spring.

The historical record of climate is relatively short; measurements of air temperature and precipitation in the Mono Basin have been made since 1951, whereas descriptions of many regional weather events date back to the mid-1800s. Other important information, such as rain and snowfall records in the nearby Sierra, is available for the period since 1924 and is maintained by LADWP. Analyses of these data can be used to define mean values of monthly air temperature and precipitation and the relative frequency of occurrence of extreme events. Of particular importance to the Mono Basin ecosystem are the occurrence of periods of prolonged drought and of episodes of heavy precipitation and climatic trends resulting from natural events or anthropogenic effects. Prehistoric climate can be inferred from geologic evidence including glacial moraines, old shorelines, alluvial deposits, and volcanic ash layers, and also from paleobotanic evidence in pack-rat middens and tree rings.

In October 1986, the surface area of the lake was approximately 69 mi^2 (Pelagos Corporation, 1987), with a maximum depth of approximately 150 ft. There are two large islands and numerous small islets in the lake. Paoha Island has an area of approximately 3 mi^2, and Negit Island is approximately 0.5 mi^2.

Because no streams drain from the lake and ions dissolved in the streamflow and groundwater seepage concentrate as the lake evaporates, Mono Lake is highly saline and alkaline. It is currently about 2.5 times as saline as the Pacific Ocean (LADWP, 1987). As is characteristic of saline lakes, Mono Lake supports a small number of productive species, mainly brine fly (*Ephydra hians*), brine shrimp (*Artemia monica*), and a few species of algae. Large numbers of migrant waterbirds use the lake and rely on the brine shrimp and brine flies for food. No fish live in the lake. (See chapter 4 for a detailed discussion of the biology of Mono Lake.)

A distinctive feature of the Mono Basin is the tufa towers, formed underwater as calcium in freshwater springs combines with carbonates in the lake water. The deposits accumulate and form picturesque towers. As the lake level has receded, the tufa towers have been exposed (see front cover).

Considerable work has been done on the geologic history of the Mono Basin (see Appendix C). In geologic terms, the basin is a tectonic depression filled with sediment. To the west of the basin are Mesozoic granites and Paleozoic metamorphics of the Sierra Nevada escarpment. To the north, east, and southeast are the Miocene volcanics of the Bodie Hills, Anchorite Hills, and Cowtrack Mountain. To the south are Quaternary volcanics, mostly rhyolitic glass, of the Mono Craters and Glass Mountain. The basin is filled with 500 to 1350 m of glacial, fluvial, lacustrine, and volcanic deposits. The ground surface comprises glacial tills and gravelly fluvial deposits. These are predominant in the west near the base of the Sierra Nevada. Sandy, windblown pumice covers areas to the north, east, and southeast of the lake.

The area became a closed basin about 3 million years ago, when a combination of faulting of the eastern scarp of the Sierra Nevada and downwarping of the north and south sides of the basin occurred. The area is still tectonically active. Mono Craters have been intermittently erupting for the last 33,000 years.

PREHISTORIC AND HISTORIC FLUCTUATIONS IN LAKE LEVEL

Mono Lake, which is approximately one-half million years old, is one of the oldest lakes in North America (Lajoie, 1968). During the last glaciation (the Wisconsin), the level of the lake was much higher than the current level of 6380 ft above sea level. This Pleistocene lake is known as Lake Russell. Fluctuations of the lake level from 23,000 through 12,500 years ago (the Tioga advance of Wisconsin glaciation) are summarized by Lajoie (1968). The lake may have reached its maximum extent 22,000 years ago, when it overflowed into the Owens Valley. At that time, the lake was 7 times deeper and 5 times larger in surface area than it is today. At its minimum during glaciation, the lake probably stood near 6600 ft, only a few hundred feet higher than the current level.

Stine (1984) has reconstructed the fluctuations in lake level over the past 3500 years using geomorphic, biotic, historic, and radiocarbon evidence along with information from sedimentary sequences and tephra exposed in the stratigraphy (Figure 1.3). The figure indicates that the lake was at its lowest about 1850 radiocarbon years before the present.

Though the lake level has been lower in prehistoric time than it has been more recently, the volume of water was probably greater than it would be for the same lake level today. Changes in the lake's morphometry resulting from volcanic activity on the lake floor during the past 600 years have altered the relationships between lake elevation and volume (S. Stine, University of California, Berkeley, personal communication, 1986). The most significant displacement of water has occurred with the formation of Paoha Island from volcanic activity and uplift of lake sediments sometime between 1723 and 1850 AD. Because the area is volcanically active, uncontrollable geologic processes could again alter the relationship between lake elevation and volume in the future.

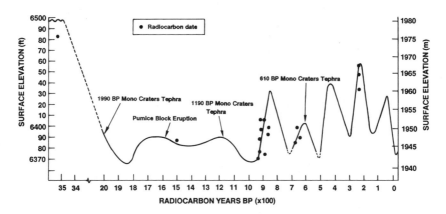

FIGURE 1.3   Surface fluctuations of Mono Lake 3500 BP to present (Stine, 1984).   (BP is relative to 1950.)

LADWP has kept records of lake levels since 1912. Lake levels have been estimated back to 1857 using cartographic, historical, and climatic evidence.   Figure 1.4 shows these estimated and measured historic lake levels.

Water has been diverted from the basin since 1941.   The level of the lake fell approximately 45 ft, from 6417 ft in 1941 to 6372 ft, the historic low stand, in 1982.   The years since 1982 have been wet, and the lake level has risen approximately 8 ft, to 6380 ft, as of August 1986.

COMPARISON OF MONO LAKE WITH OTHER SALINE LAKES

Saline lakes occur on every continent and often are the only surface water in dry regions (Hammer, 1986).   Most inland saline lakes are shallow and fluctuate widely in salinity, area, and depth.   In contrast, large, deep lakes such as Mono Lake usually experience muted salinity variation (Langbein, 1961).   However, gradual changes in depth and salinity, mainly as a consequence of climatic variations and human modification of inflows, do occur for even the largest lakes.

Among saline lakes there is a wide variety of chemical compositions and total salt contents because of the broad

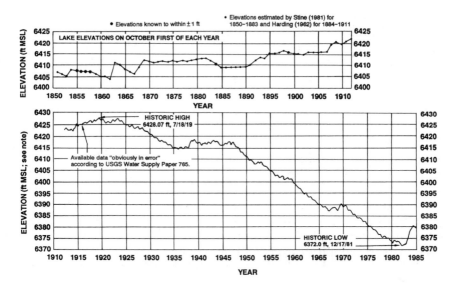

FIGURE 1.4 Surface elevations of Mono Lake 1850 to the present (Vorster, 1985). Based on Stine (1981) for estimates of levels from 1850 to 1883, Harding (1962) for 1883 to 1911, and recorded measurements for 1912 to 1984. NOTE: All lake elevations are LADWP figures in feet above sea level corrected to the USGS datum of 1929.

range of geologic and climatic settings in which these lakes occur (Eugster and Hardie, 1978; Nissenbaum, 1980; Hammer, 1986). Although the distinction is somewhat arbitrary, saline lakes are usually now recognized as those with dissolved solutes of more than 3 g/l. The upper limits of salinities depend on the concentrations at which the constituents become saturated and precipitate. The highest salinity on record (474 g/l) is that of Don Juan Pond, Antarctica, a calcium chloride water (Meyer et al., 1962).

The volume of saline water in lakes and land-locked seas is almost as large as the volume of the world's fresh water in lakes and rivers. Most of the saline water lies in the Caspian and Aral seas and in a few lakes such as Issyk Kul (USSR), Pyramid and Great Salt lakes (United States), and the Dead Sea and Lake Van (Asia Minor) (Hammer, 1986). Hence, although large, deep, saline lakes are rare, they represent a significant fraction of inland saline water. Melack (1983) selected a subset of such lakes with morphometric and chemical similarities to Mono Lake (Table 1.1) and attempted to decipher ecological characteristics that could increase understanding of Mono Lake. Only seven lakes met Melack's criteria, although additional lakes may be found in Tibet, south central USSR, and Mongolia. A comparative limnological analysis of even these seven lakes is confounded by the scant data available on some. General characteristics of these lakes include locations at moderate to high altitude in mountainous terrain and alkaline, sodium-rich waters, usually with high phosphate. Supersaturated dissolved oxygen, abundant animals and seasonally high algal populations, and low transparencies indicate that some of the lakes are productive. Few species live in these lakes; all except Mono contain fish.

The large and wide-ranging literature on saline lakes, recently reviewed by Hammer (1986), has particular relevance to Mono Lake in a few areas. Most pertinent are biogeographic studies of the tolerance of aquatic organisms to salinity, geochemical analyses of saline waters as a function of dilution and concentration, paleoecological examination of the extent and rates of changes of salinity, and investigations of vertical mixing, especially the occurrence of incomplete mixing, or meromixis.

In the context of our overall understanding of lakes, both saline and fresh, Mono Lake is of particular scientific value. Mono Lake is one of the oldest lakes in North America; it is far older than the many lakes formed within the past 15,000 years as the last continental glaciers retreated. The large quantity of carbonate dissolved in Mono Lake has prolonged the attainment of equilibrium with carbon-14 produced by the atmospheric testing of nuclear weapons and has provided an excellent opportunity to examine gas exchange between the atmosphere and lakes (Peng and Broecker, 1980). Such studies have a bearing on the extent of global warming to be expected from carbon dioxide increases associated with the burning of fossil fuels. Recent measurements showing that radionuclides can be many times more soluble in Mono Lake than in marine or fresh waters (Simpson et al., 1980, 1982; Anderson et al., 1982) may have important implications for radioactive

TABLE 1.1   Geographic, Morphometric, and Chemical Characteristics of Large, Deep Salt Lakes

Lake[a]	Location	Altitude (m)	Area (km²)	Mean depth (m)	Salinity (g/l)
Mono (1, 2)	38°00′N, 119°00′W	1942	150	19	90
Walker (3)	38°45′N, 118°40′W	1207	154	20	10.7
Qinghai Hu (4)	36°50′N, 100°10′W	3196	4635	19	12.5
Shala (5, 6)	7°30′N, 38°30′W	1567	409	89	16.8
Van (7, 8)	38°N, 43°E	1720	3600	53	22.2
Panggong Tso (9)	33°42′N, 78°45′N	4241	279	26	12.9
Karakul (10)	39°00′N, 73°30′E	3952	370	210[b]	10

[a]Numbers in parentheses indicate original sources cited in Melack (1983), as follows: (1) Mason (1967), (2) Los Angeles Department of Water and Power (personal communication, 1982), (3) Koch et al. (1979), (4) Academia Sinica (1979), (5) Morandini 1941, (6) Loffredo and Maldura (1941), (7) Gessner (1957), (8) Langbein (1961), (9) Hutchinson (1937), and (10) Ergashev (1979).
[b]Maximum depth.

NOTE: The criteria for selection of these lakes were a mean depth greater than 15 m, an area greater than 100 km², and an athalassic salinity greater than 10 g/l and less than 100 g/l.

SOURCE: Melack (1983).

waste disposal in these environments. Another consequence of the high carbonate concentrations is the formation of calcite-impregnated defluidization structures in the littoral sands around the lake. These structures, which can be mistaken for animals' burrows, show that not all tubelike formations are evidence of biological activity (Cloud and Lajoie, 1980). Of further relevance to Precambrian paleoecology is the recent speculation that the ancient sea had high alkalinity, high pH, and low calcium and magnesium concentrations, much as Mono Lake does today (Kempe and Degens, 1985).

REFERENCES

Anderson, R. F., M. P. Bacon, and P. G. Brewer. 1982. Elevated concentrations of actinides in Mono Lake. Science 216:514-516.

Cloud, P., and K. R. Lajoie. 1980. Calcite-impregnated defluidization structures in littoral sands of Mono Lake, California. Science 210:1009-1012.

Eugster, H. P., and L. A. Hardie. 1978. Saline lakes. Pp. 237-293 in Lakes: Chemistry, Geology, Physics, A. Lerman, ed. New York: Springer-Verlag.

Hammer, U. T. 1986. Saline Lake Ecosystems of the World. Monographiae Biologicae 59. Dordrecht, Netherlands: Dr W. Junk Publishers. 616 pp.

Harding, S. T. 1962. Water Supply of Mono Lake Based on Past Fluctuations. Water Resources Archives. University of California, Berkeley. Unpublished report.

Kempe, S., and E. T. Degens. 1985. An early soda ocean. Chem. Geol. 53:95-108.

Lajoie, K. R. 1968. Late Quaternary Stratigraphy and Geologic History of Mono Basin, Eastern California. Ph.D. dissertation, University of California, Berkeley. 379 pp.

Langbein, W. B. 1961. Salinity and Hydrology of Closed Lakes. U.S. Geological Survey Professional Paper 412. Washington, D.C.: U.S. Government Printing Office. 20 pp.

Los Angeles Department of Water and Power. 1987. Mono Basin Geology and Hydrology. Los Angeles, Calif.

Melack, J. M. 1983. Large, deep salt lakes: a comparative limnological analysis. Hydrobiologia 105:223-230.

Meyer, G. H., M. B. Morrow, O. Wyss, T. E. Berg, and J. L. Littlepage. 1962. Antarctica: the microbiology of an unfrozen saline pond. Science 138:1103-1104.

Nissenbaum, A., ed. 1980. Hypersaline Brines and Evaporitic Environments. Proceedings of the Bat Sheva Seminar on Saline Lakes and Natural Brines. Developments in Sedimentology 28. Amsterdam: Elsevier.

Pelagos Corporation. 1987. A Bathymetric and Geologic Survey at Mono Lake, California. Report prepared for Los Angeles Department of Water and Power. San Diego, Calif.

Peng, T.-H., and W. S. Broecker. 1980. Gas exchange rates for three closed-basin lakes. Limnol. Oceanogr. 25:789-796.

Simpson, H. J., R. M. Trier, C. R. Olsen, D. E. Hammond, A. Ege, L. Miller, and J. M. Melack. 1980. Fallout plutonium in an alkaline, saline lake. Science 207:1071-1073.

Simpson, H. J., R. M. Trier, J. R. Toggweiler, G. Mathieu, B. L. Deck, C. R. Olsen, D. E. Hammond, C. Fuller, and T. L. Ku. 1982. Radionuclides in Mono Lake, California. Science 216:512-514.

Stine, S. 1981. A Reinterpretation of the 1857 Surface Elevation of Mono Lake. California Water Resources Center Report 52. Davis, Calif.: University of California, California Water Resources Center. 41 pp.

Stine, S. 1984. Late Holocene lake level fluctuations and island volcanism at Mono Lake, California. Pp. 21-49 in Holocene Paleoclimatology and Tephrochronology East and West of the Central Sierran Crest, S. Stine, S. Wood, K. Sieh, and C. D. Miller, eds. Field Trip Guidebook for the Friends of the Pleistocene, Pacific Cell, October 12-14, 1984. Palo Alto, Calif.: Genny Smith Books.

U.S. Congress, House of Representatives. 1983. Establishing the Mono Basin National Forest Scenic Area in the State of California. House Report 98-291. Washington, D.C.: Committee on Interior and Insular Affairs. 16 pp.

Vorster, P. 1985. A Water Balance Forecast Model for Mono Lake, California. Master's thesis, California State University, Hayward. Earth Resources Monograph No. 10. San Francisco, Calif.: U.S. Forest Service, Region 5.

2
Hydrology of the Mono Basin

INTRODUCTION

Understanding the hydrology of the Mono Basin is important both as a basis for constructing water balance models to predict future lake levels and as a means for assessing potential changes in the availability and salinity of water that might affect the ecosystem of the basin. For example, lake levels directly control the position of the shallow water table around the lake and thus the availability of shallow groundwater for nearshore vegetation. Similarly, salinity and its consequences for wildlife are determined by the amount of water that flows into the lake.

This chapter gives an overview of the hydrologic processes in the Mono Basin and the models used to predict future lake levels and salinity. The discussion has four parts: (1) a general review of the meteorology and climatic influences in the region; (2) a description of hydrologic processes in the basin and a brief assessment of the available data; (3) a review of currently used water balance models for the lake or basin; and (4) a description of predicted lake levels and salinities with these models.

HYDROMETEOROLOGY

The hydrology of the Mono Basin is principally controlled by the amount and distribution of precipitation it receives, which in turn is a function of the meteorology of the Great Basin. The following discussion, providing background information on the hydrometeorology of the region, is thus based in large part on works that consider the meteorology of the Great Basin as a whole.

Synoptic-Scale Weather Systems and Air Masses

Synoptic-scale, or large-scale, weather systems are responsible for most of the precipitation around the Mono Basin, and variability in those systems causes variability in the precipitation. In his investigation of precipitation characteristics of the Great Basin, Houghton (1969) identified three principal regimes, all of which occur throughout the Great Basin. Each of these regimes is dominant in different sectors of the region and has recognized circulation patterns and air mass trajectories, each of which brings precipitation to the Mono Basin in different seasons. The Pacific component, (including polar and subtropical flows), has a winter precipitation maximum and is predominant in the western, northern, and southern sectors; the continental component, with a spring precipitation maximum, is predominant in the central and eastern sectors; and the Gulf component, with a summer precipitation maximum, is predominant in the southeastern sector. The source regions and trajectories of the air masses are shown in Figure 2.1.

Precipitation from Pacific storms, which is the majority of the precipitation in the area, is associated with ascending air in frontal zones and associated upper troughs, and with orographic lifting over the Sierra Nevada and other mountain ranges. The Sierra Nevada acts as a barrier to the moisture. Most of the precipitation from these storms falls in the high elevations, with very little reaching the east side of Mono Lake.

A substantial amount of the precipitation that falls over the Great Basin is associated with nonfrontal cyclones involving modified polar air (Houghton et al., 1975; Monteverdi, 1976). Such circulations (known in Nevada as "Tonopah Lows") are important for precipitation in eastern California in general and the Mono Basin in particular,

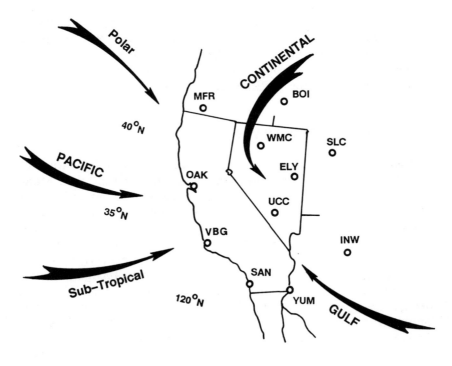

FIGURE 2.1 Major air flow patterns and air mass types affecting California and the Great Basin.

including heavy snowfall around the Owens Valley and White Mountains (LaMarche, 1974). Often these storms entrain maritime air from the Pacific west and south of southern California. The cyclone may then move slowly eastward, bringing moderate to heavy precipitation to the eastern Sierra and Mono Basin.

Summer rainfall in Arizona and New Mexico, as well as in adjacent areas of California (including the Mono Basin), Nevada, and Colorado, is mainly dependent on air mass thunderstorms or organized synoptic-scale convective storms involving air from the tropical Pacific, the Gulf of California, or the Gulf of Mexico (Hales, 1972, 1974). In the summer, solar heating of the Southwest Plateau favors the development of anticyclonic flow over the Great Basin (Reiter and Tang, 1984; Tang and Reiter, 1984). At the same time, dry Pacific air is involved in a diurnal monsoon or large-scale sea breeze across California and the Sierra. Where the westerly flow meets the cyclonic southerly or southeasterly flow of air from Arizona and Baja California, it forms a shear line or line of convergence. This shear line moves back and forth over southeastern California and western Nevada and is often coincidental with climatologically important phenomena such as lightning, blowing dust, hailstorms, and flash floods that affect Mono Lake and the surrounding mountains.

During anticyclonic conditions in winter, fog often forms over Mono Lake. With the absence of wind, this fog prevents further evaporation. The effects of cloud cover on evaporation are more complex because clouds are often accompanied by strong winds. It is conceivable that a series of wet seasons such as 1982-1983 with relatively short, cool summers could account for a decrease in evaporation, contributing to a rapid increase in size of Mono Lake and other Great Basin lakes.

Precipitation Patterns Related to the Fall and Rise of Great Basin Lakes

During the most recent years of the historical period (1975 to 1986), two related hydrometeorological phenomena have attracted attention: the increased incidence of extreme weather events, including extremely wet and extremely dry periods (Policansky, 1977; Goodridge, 1981; Karl et al., 1984) and the rise in level of Great Salt Lake, Pyramid Lake, Walker Lake, and Mono Lake.

To review the temporal and spatial extent of both droughts and periods of greater than normal precipitation in the Great Basin, the average monthly and annual precipitation at eight locations (Elko, Ely, Las Vegas, Reno, and Winnemucca in Nevada; Milford and Salt Lake City in Utah;

and Bishop, California; see Figure 2.2) for the period 1951 to 1980 were analyzed (H. Klieforth, University of Nevada, personal communication, 1986). The elevations range from 2162 ft at Las Vegas to 6253 ft at Ely. At Ely, Milford, and Salt Lake City, spring is usually the wettest season; at Las Vegas, winter and summer are the wettest; and for the other four, winter is the wettest season.

In the historical record of extreme precipitation events, the contrasting 2-year periods of 1975 to 1977 and 1981 to 1983 are outstanding. The former years were extremely dry and the latter extremely wet. In 100 years of rainfall records in California, including the Sierra Nevada, the occurrences of two consecutive extremely dry years and two consecutive extremely wet years were unprecedented (Goodridge, 1981).

The rise in Great Basin lakes during the 1980s is well documented. In June 1986 Pyramid Lake had risen to an elevation of 3817 ft above sea level, higher than it had been since 1944, and Mono Lake had risen 8 ft since 1982. While the levels of Mono and Pyramid lakes and the flow of the Truckee River are affected by releases from dams upstream and by various diversions for irrigation, there is nevertheless a close correlation between their recorded levels and the record of precipitation, particularly that at higher elevation.

It is apparent from the recent historical record that transitions from dry to wet regimes and back are relatively abrupt. These shifts may be related to preferred wavelengths in the upper air flow and these in turn to major physiographic features and to variable thermal influences such as sea surface temperatures.

During the recent extreme events of the 1980s, a search for causes focused attention on the El Niño-Southern Oscillation (ENSO) phenomenon (Kiladis and Diaz, 1986). A strong ENSO development leads to extreme events of opposite character in various parts of the world, including devastating droughts in some regions and excessive rainfall in other regions. There were El Niño events in both 1976-1977 (dry in California) and in 1982-1983 (wet in California) (Ramage, 1986). An additional likely cause of climate change is the global warming expected to result from in-

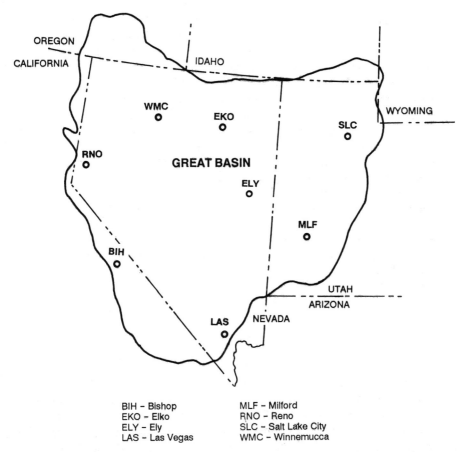

FIGURE 2.2 Great Basin region and eight weather stations selected for precipitation study.

creased amounts of carbon dioxide and other spectrally active trace gases in the atmosphere.

HYDROLOGIC PROCESSES

The most basic concept in hydrology is the hydrologic cycle--the continuous transfer of water between the surface (e.g., oceans and lakes), the atmosphere, and the subsurface (i.e., groundwater). Water balance models attempt to quantify components of the hydrologic cycle for a specified region using conservation of mass for inflows, outflows, and changes in storage.

Mono Basin is a closed basin, the only natural outflows being evaporation from the lake and soil and evapotranspiration from the sparse vegetation. Moisture input to the basin occurs as snow and rainfall. Water derived from melting snow and rainstorms reaches streams by overland flow and groundwater seepage. Thus, the dominant processes controlling the distribution of water within the basin are precipitation, surface runoff in streams, groundwater discharge to the lake, lake evaporation, and terrestrial evapotranspiration. The following sections describe these processes in more detail and discuss data that are available to estimate each as a component of the moisture budget for the basin.

Precipitation

The average annual precipitation in the Mono Basin varies from about 6 in. at the east side of the lake to about 50 in. at higher elevations in the Sierra Nevada. Although intense, localized thunderstorms occur in the summer, the greatest amount of precipitation falls in the winter. Approximately 75 percent of the annual precipitation occurs between October and March (Vorster, 1985).

The Sierra snowpack is the principal source of surface runoff in the basin. Snowfall occurs year-round at high elevations and begins to accumulate in middle to late October. Snowmelt begins in April and continues through May, with maximum amounts in May and June. Vorster (1985) suggests that about 77 percent of the average annual precipitation at elevations above 8500 ft is contained in the snowpack on April 1. This is not an unreasonable assumption, but in years of particularly heavy summer and fall rains the nonsnow precipitation may contribute more than 23 percent of the annual precipitation.

Although few precipitation gages with continuous records are present in the basin, mean annual precipitation appears to be adequately known relative to other hydro-logic components. Table 2.1 shows average precipitation for the eight precipitation stations in the basin. Locations of the stations are shown in Figure 2.3.

In addition to rainfall measurements, snow-course data are available from nine locations in or adjacent to the basin. A summary of rainfall at gaging stations and iso-hyetal maps of mean annual precipitation over the basin are given by Vorster (1985) and LADWP (1987). The positions of the isohyetals differ, particularly at high elevations where Vorster utilized snow-course data and in the eastern side of the basin where few gaging stations are located. Nevertheless, both studies estimate the mean annual precipitation over the lake to be approximately 8 in.

Surface Runoff

Much of the surface runoff in the basin originates in the Sierra Nevada, where most precipitation occurs. Five major Sierra Nevada streams (Rush, Lee Vining, Mill, Walker, and Parker), as well as a number of smaller streams, drain into the basin from the Sierra and other surrounding hills. Because runoff from the Sierra is fed primarily by snowmelt, streamflows are highly seasonal, with one-half to two-thirds of the total annual flow occurring in May, June, and July (Vorster, 1985).

Locations and descriptions of surface runoff gaging stations are given in detail by both Vorster (1985) and LADWP (1987). The total mean gaged surface runoff in the principal Sierra streams is approximately 150,000 acre-ft/yr. This represents about 75 to 85 percent of total surface and subsurface inflows to the basin. Approximately 75 percent of the gaged runoff is measured on the two largest streams, Rush and Lee Vining creeks. LADWP estimated total unmeasured flows to be about 25,000 acre-ft/yr, while Vorster (1985) gives a higher estimate as the sum of two components, unmeasured Sierra runoff of approximately 17,000 acre-ft/yr, and non-Sierran runoff of about 20,000 acre-ft/yr. The most significant creeks that are not gaged and reported on a regular basis are Wilson, Bridgeport, Cottonwood, and Post Office creeks.

TABLE 2.1 Average Annual Precipitation at Gaging Stations in Mono Basin (LADWP, 1987)

Station	Period of Record	Elevation (ft)	Average Precipitation (in.)	
			Period of Record	Period of 1941-1985
Bodie	1965-1968	8370	19.2	--
Cain Ranch	1931-1932 to 1984-1985	6850	11.44	11.34
East Side Mono Lake	1975-1976 to 1984-1985	6840	5.70	--
Ellery Lake	1925-1926 to 1984-1985	9645	25.68	20.42
Gem Lake	1925-1926 to 1984-1985	8970	21.81	20.91
Mark Twain Camp	1950-1955	7230	6.80	--
Mono Lake	1951-1968	6450	12.50	--
Rush Creek Power House	1957-1979	7235	25.20	--

Estimates of ungaged runoff will clearly introduce inaccuracies into computation of the moisture budget for the basin. Another source of error in the calculations is the fact that the actual surface runoff into the lake is unknown. Most stream gaging stations are located above LADWP diversion points, 4 to 8 mi from the lake. Releases past these points flow over porous channel beds to underlying aquifers. Thus surface runoff inflows to the lake, distinct from groundwater inflows, cannot be estimated accurately.

Groundwater Occurrence and Movement

The movement of groundwater is strongly controlled by the geology of the basin. As discussed in chapter 1, the basin is filled with layers of interfingered glacial, fluvial, lacustrine, and volcanic deposits. These basin sediments form a complex series of confined and semiconfined aquifers and aquitards, which are recharged by precipitation in adjacent hill and mountain areas. The water table is generally within 50 m of the ground surface throughout the

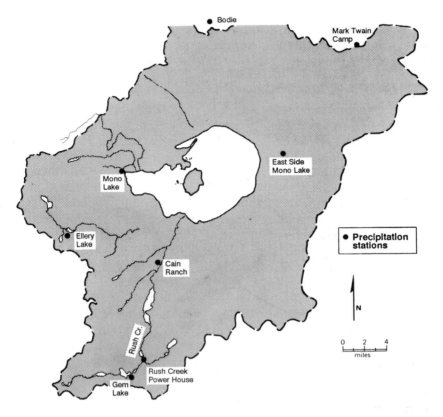

FIGURE 2.3 Locations of precipitation gaging stations in the Mono Basin (LADWP, 1987).

basin and occurs at much shallower depths near the lakeshore (Lee, 1969). Mono Lake acts as a regional groundwater sink; groundwater moves toward the lake, discharging at discrete springs and zones of diffuse seepage along the lakeshore and beneath the lake.

Nearshore Groundwater Flow

From the perspective of the Mono Basin ecosystem, the most important aspect of groundwater flow is the seepage at the lake-sediment interface and in nearshore marshes and salt flats. Several studies have described the locations of springs and shallow groundwater gradients around the lake (Lee, 1969; Loeffler, 1977; Vorster, 1985; LADWP, 1987). Locations of the largest springs, as mapped by LADWP (1987), are shown in Figure 2.4. Zones of seepage on the north and east sides of the lake can be seen on infrared photos as strips of vegetation (see back cover).

To supplement and update the information in previous reports, this committee, in conjunction with LADWP, installed 23 shallow piezometers along four transects around the lake in September 1986 (see Figure 2.5). At each location at least one test hole was drilled by LADWP to a depth of more than 20 ft using a water jet rig. With this method, a small-diameter steel casing, slotted over the lower 5 ft, was installed during drilling. Other shallow test holes were hand augered and cased with 1.25-in.-diameter PVC that was capped at the bottom and slotted over the lower 4 to 6 in. of casing. Each hole was filled with sand around the slotted section, and then backfilled to the ground surface with sediments from the augered hole. Water levels and specific conductivity were measured about 3 to 7 days after installation when water levels in the test holes had equilibrated. LADWP has monitored water levels in the wells at monthly intervals since the installation. Results from these test holes are summarized in Table 2.2.

Shallow groundwater gradients to the lake, rates of groundwater flow, and springflow chemistry vary greatly around the lake. Gradients are highest and total dissolved solids lowest at the west side of the lake, where groundwater inflow from the Sierra Nevada is greatest. In contrast, gradients are very low on the north and northeast side, where little groundwater inflow occurs. Nearshore shallow groundwater in this area has high total dissolved solids and is either residual lake water, remaining after the lake elevation declined, or lake water drawn into the sediments by evaporation in the salt flats. In areas of extensive tufa development, such as the Old Marina, Simon's Spring, and Warm Springs, gradients are erratic and the locations of springs are controlled by fractures and tufa ridges.

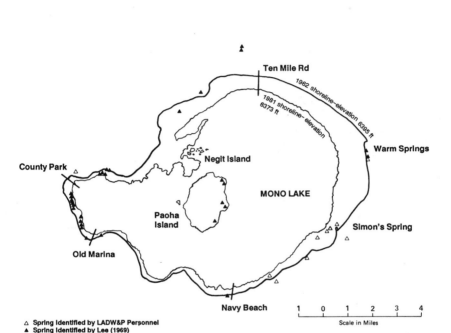

FIGURE 2.4 Locations of springs in the Mono Basin (LADWP, 1987) and locations of transects for sampling groundwater circulation.

Groundwater Data Assessment

Quantitative estimates of changes in groundwater storage and rates of movement are commonly made with
numerical groundwater flow models. Construction of such
models requires a description of the geometry and hydraulic
properties of geologic deposits as well as measurements of
water table elevations or hydraulic heads in wells distributed over the study area.

Information on the composition of the basin fill, and
thus the aquifer geometries, hydraulic properties, and hydraulic heads, is extremely sparse. According to Lee

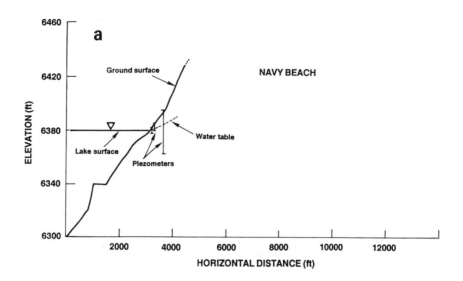

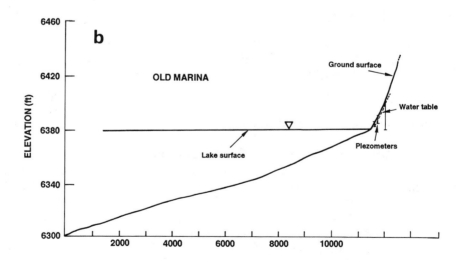

FIGURE 2.5   Location of test holes at transects.

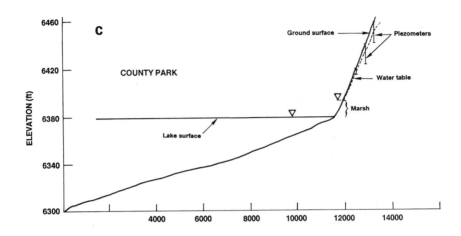

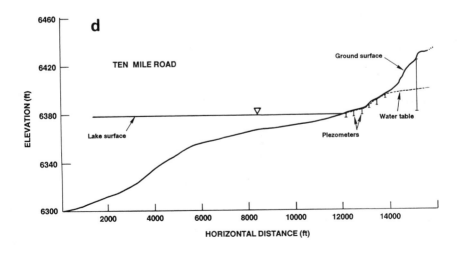

FIGURE 2.5   (continued)

TABLE 2.2 Test Hole Measurements

Location	Test Hole No.	Ground Elevation (ft)	Depth of Hole (ft)	Temperature (°C)	Specific Conductivity (μmhos)	Depth to Water (ft) 9/16	10/31	11/20
Simon's Spring	D-1A/634T	6418.6	37.50	17.0	48	3.80	3.70	3.60
Navy Beach	D-2A/635T	6395.1	31.60	too deep	too deep	3.25	8.04	10.10
	D-2B	6384.5	4.95	35.0	3950	3.50	3.89	3.90
	D-2C	6382.1	3.25	31.0	2000	1.30	1.85	1.90
Old Marina	D-3A/636T	6402.6	20.7	11.3	173	flowing	0.03	0.02
	D-3B	6394.9	4.2	13.0	720	1.8	0.66	0.61
	D-3C	6389.4	5.9	15.5	510	3.0	0.45	0.41
	D-3D	6386.2	4.7	15.7	980	3.2	0.60	0.51
County Park	D-4A/639T	6457.8	20.3	17.5	52	6.8	9.13	10.2
	D-4B/636T	6440.9	17.0	17.0	50	3.2	4.81	5.1
	D-4C	6421.0	4.5	12.5	388	1.7	1.33	1.3
	D-4E	6395.1	1.5	13.8	372	0.3	0.35	0.3
10-Mile Road	D-5A/639T	6426.4	42.0	too deep	too deep	19.80	19.75	19.70
	D-5B	6399.1	2.0	14.7	23,500	1.30	0.80	0.61
	D-5C	6392.6	4.2	16.5	9,300	1.10	0.00	0.00
	D-5D	6389.7	3.1	16.8	17,500	1.30	1.22	1.06
	D-5E	6385.2	4.7	17.0	15,300	0.43	0.03	0.00
	D-5F	6383.6	3.8	16.0	29,500	1.10	1.27	1.20
	D-5G	6381.5	4.6	17.5	28,300	1.20	1.23	1.24
	D-5B-1	6399.4	18.40	plugged	plugged		0.37	0.31
	D-5B-2	6399.4	8.00	15.0	7,000	0.36	0.02	0.01
	D-5B-3	6399.4	3.28	14.7	6,500	0.43	0.02	0.02
Warm Springs	D-6A/640T	6419.3	41.5	24.0	162	2.70	3.18	3.20

(1969), few lithologic logs were available for wells in the basin prior to 1969. With the exception of shallow domestic wells, only two water supply wells, Cain Ranch and Lundy Land Company, were in use at that time. Two deep oil exploration wells were drilled on the west side of the basin in 1908 and 1911 (Lee, 1969). In 1971 two deep geothermal wells were drilled, one on the south shore near Panum Crater and a second on the north shore east of Black Point (LADWP, 1987). LADWP installed a test well 256 ft deep near Lee Vining Creek in 1980. In addition, several domestic water supply wells have been drilled recently on the west side of the basin.

On the basis of the few available lithologic descriptions from wells, some general observations about the physical characteristics of the sediments in the western portion of the basin can be made. West of the lake, glacial fluvial sediments with high permeability interfinger with a complex sequence of volcanic ash and lacustrine deposits with relatively low permeability. These layers form a vertical series of confined aquifers with a shallow unconfined aquifer in fluvial sediments above the uppermost silt and ash deposits. Glacial sediments are assumed to pinch out as lacustrine deposits thicken toward the center of the basin. Near Lee Vining Creek, at LADWP well 373, the upper unconfined aquifer is composed of fluvial gravels that extend to a depth of about 6 m. The aquifer is underlain by low-permeability, lacustrine clayey silts and rhyolitic ash layers. Logs from a 609 m-deep well, drilled on Paoha Island in 1908, indicate that the well penetrated at least three confined aquifers. The Dechambeau well, drilled at Black Point in 1911, apparently intersected eight separate confined aquifers (Lee, 1969). The water temperature in these aquifers increased progressively with depth.

No data are available on the sediment composition or hydraulic properties of fill elsewhere in the basin. Thus, the geometry, areal extent, and hydraulic conductivity and storativity of individual aquifers are currently unknown. This lack of control on aquifer geometries and hydraulic properties is the fundamental reason why no attempt has been made to construct a numerical groundwater model to estimate rates of groundwater flow into the lake.

Lake Evaporation

Evaporation from Mono Lake is a complex process dependent upon solar radiation, vapor pressure gradients, lake and air temperature, wind, lake salinity and surface area, and wave action. Vorster (1985) summarizes the results of 17 evaporation studies at Mono Lake. These estimates, made between 1934 and 1985, range from a high of 78.8 in./yr to a low of 37.4 in./yr, with a mean of approximately 46 in./yr. LADWP (1987) estimated the annual freshwater evaporation rate to be 42 in./yr.

Evaporation is one of the most poorly quantified hydrologic components in the basin. Although evaporation is approximately 50 percent (LADWP, 1987) of the total water loss, including exports, from the basin, most estimates are based on inaccurate evaporation pan measurements or empirical evaporation-altitude relationships for the eastern Sierra. Given the importance of lake evaporation in the basin water balance, additional efforts should be made to obtain better estimates of evaporation from the lake.

Terrestrial Evapotranspiration

Terrestrial evapotranspiration in the Mono Basin occurs as evaporation from the bare ground, evaporation of subsurface soil moisture, and evapotranspiration from vegetation. Terrestrial evapotranspiration can be estimated from direct measurement with lysimeters, with moisture flux methods using measurements of soil moisture change over time, or with empirical moisture budget methods. These latter methods, including the Blaney-Criddle (Blaney, 1954) and Thornthwaite and Mather (1957) methods, are generally applied when detailed measurements are not available.

Vorster (1985) estimated rates of evapotranspiration from riparian vegetation, irrigated land, and nearshore phreatophytes using the Blaney-Criddle method. Vorster estimated that between 1937 and 1983 an average of about 18,000 acre-ft/yr were lost from the Mono Basin, excluding xerophyte losses. This represents about 7 percent of total basin outflows including exports from the basin. Thus, terrestrial evapotranspiration, excluding that from xerophytes, which utilize local precipitation, appears to be a relatively minor component in the basin moisture budget.

DESCRIPTION AND ASSESSMENT OF WATER BALANCE MODELS

Although the water balance approach to hydrologic studies is conceptually simple, accurate water balance models are difficult to construct. Measurements of the components are rarely complete, and measurement errors may be large. The instrumentation required to adequately describe the individual components of such a model is extensive. Very few watersheds in the United States are monitored in sufficient detail to describe all the processes. For this reason, water balance models are generally designed to derive maximum information from the available data. This is the case for past and present models for Mono Lake and the Mono Basin.

Vorster (1985) surveyed and evaluated previously developed water balance models for the Mono Lake and Basin. In addition, Todd (1984) reviewed the two most recent models by Vorster and LADWP in the context of previous studies. Because these reviews are available and because Vorster's model (1985) and LADWP's model (1984 and 1987) are the most extensive and complete, the following discussion is limited to these two models. Vorster's model was developed as a master's thesis in geography at the California State University, Hayward. The LADWP model, first published in November 1984, has been revised and updated as more data have become available. The most recent version was published in January 1987.

The basic equation for a hydrologic mass balance model states the conservation of mass over a specified region:

$$I - O = \Delta S + E_R$$

where I represents inflows to the solution domain, O is outflows from the domain, ΔS is the change in storage, and E_R represents residual errors due to measurement errors and inaccuracies and unknown or unmeasured components. While a complete model might include all of the components shown in Figure 2.6, often the decision of which individual terms are included in inflow and outflow estimates is determined by available data. Selection of the solution domain or the boundaries of the study region is of fundamental importance because it also affects components that must be defined in order to estimate inflows and outflows. Thus data availability as well as physical setting should be considered in the selection of the study region boundary.

In the case of Mono Lake, at least three choices of boundary location and moisture balance equation are possible (Table 2.3). The solution of any one of these equations would provide an estimate of changing lake volumes or elevations if the other terms in the moisture balance equation can be estimated accurately. Selection of the most appropriate boundary to use is determined by which equation can be solved most accurately using available data.

For example, Case I represents a traditional approach to hydrologic water balance models. Here surface water inflows need not be estimated because the choice of the problem domain coincides with surface water drainage divides. On the other hand, this treatment requires quantification of snowmelt and snowpack storage, as well as of evapotranspiration and changes in groundwater storage, components that are difficult to measure accurately.

In Case II, the study region boundary is located at the contact between unconsolidated basin fill sediments and lower permeability glacial tills and rocks of the Sierra. This boundary was used by Vorster (1985), who further expanded the moisture balance equation to include 18 terms. The major advantage to this approach is that inflows to the system are relatively well-defined. Assuming groundwater inflows into the valley sediments are small, inflow is defined by precipitation on the lake and basin fill and streamflows across the boundary of the problem domain where approximately 75 to 85 percent of the surface runoff is gaged. Vorster did not treat groundwater inflow as a distinct term, but instead combined unknown groundwater inflows and unmeasured surface runoff inflows into a single term. Losses, due to export, lake evaporation, and

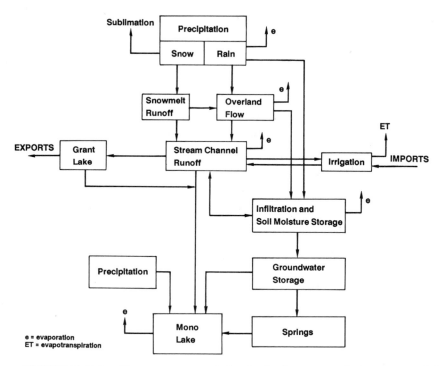

FIGURE 2.6 Components of hydrologic mass balance model for Mono Lake.

terrestrial evapotranspiration, are less well-known than inputs, but generally must be specified in any water budget model. Changes in storage occur as changes in soil moisture storage, surface runoff storage, groundwater storage, and lake storage. For an annual time interval, change in lake storage is probably the most significant of these and is relatively well-known from historic measurements of lake levels.

Case III, used by LADWP (1984, 1987), considers the water budget of the lake only. This is the most straightforward approach to a moisture budget model for predicting lake levels in the sense that it includes the smallest number of components. However, several of these components are poorly known. Inflows include surface flow into the lake, groundwater inflows, and precipitation on the lake.

TABLE 2.3   Moisture Balance Equations for Mono Lake for
Three Possible Solution Domains

Case I: A boundary that encompasses the entire Mono Basin watershed. The governing
equation is:

$$PW + IM - EX - EWL - EML - EWT = \Delta SML + \Delta SWLS + \Delta SWGW + \Delta SWSM + \Delta SWSR$$

where: PW = precipitation on the entire Mono Basin
         IM = imports to the Mono Basin
         EX = exports from the Mono Basin including groundwater seepage into the
              LADWP tunnel
         EWL = evaporation from all of the lakes within the basin excluding Mono
              Lake
         EML = evaporation from Mono Lake
         EWT = terrestrial evapotranspiration
         ΔSML = change in storage of Mono Lake
         $\Delta SWLS$ = change in storage of all watershed lakes
              except Mono Lake
         $\Delta SWGW$ = change in storage of groundwater
              throughout the watershed
         $\Delta SWSM$ = change in storage in soil moisture
              throughout the watershed
         $\Delta SWSR$ = change in storage of surface water
              runoff

Case II: A boundary that includes Mono Lake and the surrounding groundwater storage
area (basin fill). The governing equation is:

$$PB + ISWF + IGWF - EBX - EML - EBT = \Delta SML + \Delta SBGW + \Delta SBSM + \Delta SBSR$$

where previously undefined components are:

         PB = precipitation on the basin fill and Mono Lake
         ISWF = surface water inflows to the basin fill
         IGWF = groundwater inflows to the basin fill
         EBX = exports from study region
         EBT = evapotranspiration from basin fill including
              losses from lakeshore vegetation
         $\Delta SBGW$ = change in groundwater storage in the basin fill
         $\Delta SBSM$ = change in soil moisture storage in the basin fill
         $\Delta SBSR$ = change in storage of surface runoff within the basin fill

Case III: A boundary at the lakeshore. Study region includes only Mono Lake. The
governing equation is:

$$PL + ISWL + IGWL - EML = \Delta SML$$

where previously undefined components are:

         PL = precipitation on Mono Lake
         ISWL = surface water inflow to Mono Lake
         IGWL = groundwater inflow to Mono Lake

Because the stream gaging stations are located several miles from the lakeshore, both surface water and groundwater inflows are unknown and must be lumped with the residual error term in the model. Outflows are due entirely to evaporation from the lake surface, a poorly measured process. The change in storage is the relatively well-known change in lake storage estimated from lake level measurements.

Vorster's model is the most detailed model that has been developed for the Mono Lake or Basin and is constructed to take advantage of available hydrologic data. However, the reliability and accuracy of all current hydrologic models are limited by poor estimates of major components of the hydrologic cycle. Models by both Vorster and LADWP require estimates of mean annual precipitation over the lake, surface and subsurface inflows, and lake evaporation.

Estimated values for these parameters are highly variable. For example, estimates of the average annual precipitation rate on Mono Lake range from 5.3 to 12.0 in./yr. Estimates of average annual inflows from ungaged watersheds vary from 0 to 113,000 acre-ft/yr, and lake evaporation estimates range between 37.4 to 78.8 in./yr (Vorster, 1985).

Vorster performed a sensitivity analysis to assess the effect of uncertainty in the data on the ability of his model to predict observed lake levels. As expected, uncertainty in lake evaporation estimates had the greatest influence on model results. Depending on the volume of surface water exports, a change of ±5 percent in estimates of lake evaporation rates resulted in a variation of 2 to 14 ft in projected long-term lake levels. No error analysis has been performed on the LADWP model.

Limitations in available data are the major large source of error in the moisture budget models. Of most importance is the need for more accurate measurements of lake evaporation. In addition, estimation of groundwater inflows to the lake, monitoring of major ungaged streams, and measurements of precipitation on the east side of the lake would improve the reliability and accuracy of the model simulations.

MODELING OF MONO LAKE LEVELS AND SALINITY

For the purposes of this report, two relationships predicted by the hydrological models need to be determined:
(1) the relationship between elevations of Mono Lake and
streamflow at the points of the diversions and (2) the relationship between lake elevation and salinity. The former is
required for describing the effects of changes in lake level
on the riparian (stream) systems. The latter is required
for predicting the lake levels at which the aquatic biota
will be affected by increased salinity (see chapter 6).

The two available models of the hydrology of the Mono
Basin (Vorster, 1985; LADWP, 1987) were modified for this
report by using synthetically generated sequences of
streamflows as inputs rather than historical streamflows.
The historical streamflows of approximately 40 years' duration were extended to 2000 years by using an autoregres-
sive moving average model (Box and Jenkins, 1970). Synthetic streamflows were used because they preserve the
statistical properties of the original record (mean, variance,
and skewness) while allowing equally likely hydrologic
events to be input to the models. This approach minimizes
the cyclical modeling results of the brief 40-year historical
data set used in the Vorster and LADWP models. The generation of 2000 years of data is arbitrary but considered
sufficient for the required modeling results.

For standardization, an evaporation rate of 42 in./yr
was used in both models. The LADWP and Vorster models
use different values of freshwater evaporation as inputs to
their models. The former uses 40.8 in./yr, and the latter
uses 45 in./yr. Because of the large uncertainties in the
evaporation data, and in order to eliminate an excessive
number of modeling results, an approximate freshwater rate
of 42 in./yr was used for this report as an input value for
both models. This annual freshwater evaporation rate is
converted to an annual saline water evaporation rate independently as a function of lake volume and specific gravity
by each model. Vorster's model was also adjusted to include the most recent bathymetric data from Pelagos Corporation (1987). Both models were then used to simulate
the elevation of Mono Lake as a function of flow at the
diversion points (Figure 2.7).

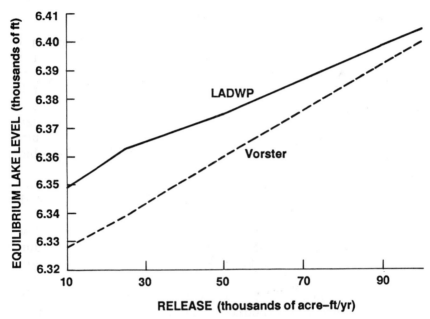

FIGURE 2.7 Calculations of equilibrium lake level versus flow at diversion points for an evaporation rate of 42 in./yr.

For any given amount of water exported from the basin, the lake will not attain its equilibrium level (level at which inflow of water equals outflow from evaporation) for more than 100 years. Figures 2.8 and 2.9 show the results of the Vorster and LADWP models for predicting lake level as a function of time for releases (controlled releases past LADWP diversion structures on Rush, Parker, Walker, and Lee Vining creeks) to Mono Lake of 10,000, 25,000, 50,000, 75,000, and 100,000 acre-ft/yr. In both these figures, the early fluctuations (before 200 years) are due to oscillations in the climatic data used as input to the models. After approximately 200 years the lake levels stabilize. Small fluctuations in lake level still occur due to oscillation in the climatic data.

A comparison of the results of Figures 2.7, 2.8, and 2.9 indicates that the Vorster model predicts lower lake elevations for the same values of releases to Mono Lake. For

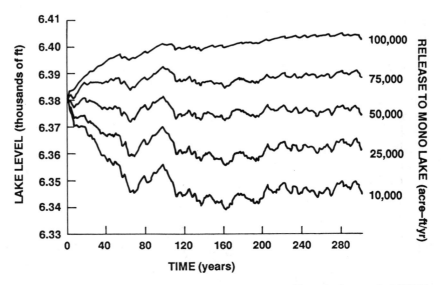

FIGURE 2.8   Lake level versus year predicted from LADWP model (1987) using an evaporation rate of 42 in./yr.

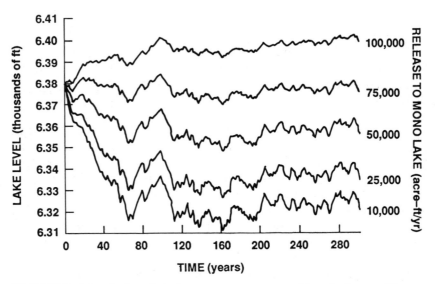

FIGURE 2.9   Lake level versus year predicted from Vorster model (1985) using an evaporation rate of 42 in./yr.

example, for a release of 10,000 acre-ft/yr, the LADWP model predicts a lake elevation of approximately 6349 ft while the Vorster model predicts a lake elevation of approximately 6328 ft. This difference of 21 ft is the largest that occurs. For a release of 100,000 acre-ft/yr, the Vorster model predicts a value that is only 4 ft below that predicted by the LADWP model. There are several reasons for these differences. Given that the two models use different boundary locations, time bases for model calibration, procedures for terrestrial evapotranspiration, and estimates of ungaged surface water runoff, the differences in the results of the two models are not unexpected.

To improve the modeling capability for the Mono Basin, a new set of models with a monthly time increment, based on a comprehensive surface water and groundwater hydrologic data collection network, is needed. This data collection network would need to focus on the components with limited or missing data, such as nearshore and deep groundwater, the ungaged surface runoff areas, lake evaporation, and terrestrial evapotranspiration.

Both Vorster (1985) and LADWP (1987) calculated the relationship between salinity and lake level by assuming a constant amount of salt in the lake. Therefore, salinity is taken to be a linear function of lake volume. The actual value used by LADWP (1987) for the total salt content of the lake is an average of summations of the major solutes analyzed separately in 11 surface samples obtained from 1940 to 1980. This number (285×10^6 tons) is similar to the average of gravimetric determinations made for several stations and depths in 1982 (288×10^6 tons) (B. White, Los Angeles Department of Water and Power, personal communication, 1987).

The assumption of a constant amount of salts in the lake appears justified for salinities below approximately 125 g/l. Above this salinity, minerals will begin to precipitate and will remove some ions from solution, as discussed in chapter 3. However, the geochemistry is not well enough understood to precisely estimate the relationship between lake level and salinity for salinities above 125 g/l. The committee adopted, for this report, the values calculated by LADWP (1986), recognizing that values for salinity above 125 g/l (corresponding to a lake level of approximately

6360 ft above sea level as discussed in chapter 6) may be overestimated.

REFERENCES

Blaney, H. 1954. Consumptive-use requirements for water. Agric. Eng. 35:870-873, 880.

Box, E. P., and G. M. Jenkins. 1970. Time Series Analysis: Forecasting and Control. San Francisco, Calif.: Holden-Day. 553 pp.

Goodridge, J. D. 1981. California Rainfall Summary: Monthly Total Precipitation 1849-1979. Sacramento, Calif.: California Department of Water Resources, Division of Planning. 55 pp.

Hales, J. E., Jr. 1972. Surges of maritime tropical air northward over the Gulf of California. Mon. Weather Rev. 100:298-306.

Hales, J. E., Jr. 1974. Southwestern United States summer monsoon source--Gulf of Mexico or Pacific Ocean? J. Appl. Meteorol. 13:331-342.

Houghton, J. G. 1969. Characteristics of Rainfall in the Great Basin. Ph.D. dissertation, University of Oregon, Eugene. 292 pp.

Houghton, J. G., C. M. Sakamoto, and R. O. Gifford. 1975. Nevada's Weather and Climate. Reno, Nev.: Nevada Bureau of Mines and Geology and University of Nevada. 78 pp.

Karl, T. R., R. E. Livezey, and E. S. Epstein. 1984. Recent unusual mean winter temperatures across the contiguous United States. Bull. Am. Meteorol. Soc. 65:1302-1309.

Kiladis, G. N., and H. F. Diaz. 1986. An analysis of the 1977-78 ENSO episode and comparison with 1982-83. Mon. Weather Rev. 114:1035-1047.

LaMarche, V. C., Jr. 1974. Paleoclimatic inferences from long tree-ring records. Science 183:1043-1048.

Lee, K. 1969. Infrared Exploration for Shoreline Springs: A Contribution to the Hydrogeology of Mono Basin, California. Ph.D. dissertation, Stanford University. 216 pp.

Loeffler, R. M. 1977. Geology and hydrology. Pp. 6-38 in An Ecological Study of Mono Lake, California, D. W.

Winkler, ed. Institute of Ecology Publication No. 12. Davis, Calif.: University of California, Institute of Ecology.

Los Angeles Department of Water and Power. 1984. Background Report on Geology and Hydrology of Mono Basin. Report of the Aqueduct Division, Hydrology Section. Los Angeles, Calif.

Los Angeles Department of Water and Power. 1986. Report on Mono Lake Salinity. Los Angeles, Calif.

Los Angeles Department of Water and Power. 1987. Mono Basin Geology and Hydrology. Los Angeles, Calif.

Monteverdi, J. P. 1976. The single air mass disturbance and precipitation characteristics at San Francisco. Mon. Weather Rev. 104:1289-1296.

Pelagos Corporation. 1987. A Bathymetric and Geologic Survey at Mono Lake, California. Report prepared for Los Angeles Department of Water and Power. San Diego, Calif.

Policansky, D. 1977. The winter of 1976-77 and the prediction of unlikely weather. Bull. Am. Meteorol. Soc. 58:1073-74.

Ramage, C. S. 1986. El Niño. Sci. Am. 254(6):76-83.

Reiter, E. R., and M. Tang. 1984. Plateau effects on diurnal circulation patterns. Mon. Weather Rev. 112:638-651.

Tang, M., and E. R. Reiter. 1984. Plateau monsoons of the northern hemisphere: a comparison between North America and Tibet. Mon. Weather Rev. 112:617-637.

Thornthwaite, C. W., and J. R. Mather. 1957. Instructions and Tables for Computing Potential Evapotranspiration and the Water Balance. Publications in Climatology 10(3). Centerton, N.J.: Drexel Institute of Technology, Laboratory for Climatology.

Todd, D. K. 1984. The Hydrology of Mono Lake: A Compilation of Basic Data Developed for *State of California* v. *United States*, Civil No. S-80-696, U.S.D.C., E.D. Cal. Berkeley, Calif.: David Keith Todd Consulting Engineers.

Vorster, P. 1985. A Water Balance Forecast Model for Mono Lake, California. Master's thesis, California State University, Hayward. Earth Resources Monograph No. 10. San Francisco, Calif.: U.S. Forest Service, Region 5.

3
Physical and Chemical Lake System

INTRODUCTION

The physical and chemical properties of Mono Lake determine the environment in which its biota live. As changing lake levels change that environment, the biota will be affected accordingly. Solar radiation provides energy for photosynthetic organisms and heats the water. Density, which is a function of temperature and concentrations of dissolved and particulate matter, determines in part the stratification or layering of water within a lake. This stratification in turn affects the amount of nutrients and dissolved gases available to organisms in the lake.

In alkaline, saline lakes such as Mono Lake, the concentration and relative proportions of the major ions (sodium, potassium, calcium, magnesium, sulfate, carbonate plus bicarbonate, and chloride) determine the osmotic environment for the organisms and the acid-base balance. Processes such as inputs of chemical constituents from surface waters and groundwater and loss of constituents through sedimentation and precipitation of minerals control the chemistry of the lake. The availability of nutrients (e.g., nitrogen and phosphorus), which enter the surface layer from the atmosphere, inflowing streams, and the more dense bottom layers of the lake, is also critical to the ecosystem.

PHYSICAL SYSTEM

The physical conditions of a lake are determined primarily by the shape of its basin, the transparency of the lake water, density variation in and motions of the water, and the local meteorology. Geologic processes that formed the basin set limits for the lake's morphometry. Changes in inputs of water or sediments caused by natural or anthropogenically induced conditions can modify morphometric features such as depth or island extent and can effect significant ecological changes. A morphometric description of Mono Lake was derived by Mason (1967) from Scholl et al.'s (1967) bathymetry and USGS topographic maps for a lake level of 6391.2 ft, the mid-July 1964 elevation. Recent bathymetric data (Pelagos Corporation, 1987) permit improved calculation of the lake's hypsographic curve and related morphometric parameters (Figures 3.1 and 3.2).

Transparency of a lake depends upon the quantity and optical properties of the materials dissolved and suspended in the water. The resultant depth of penetration of solar radiation determines the depth to which primary productivity can occur and influences the distribution of heat and hence the density of the water. The transparency of Mono Lake as estimated by Secchi disk visibility varies from a winter low of 0.5 to 1 m to a summer high of 8 to 12 m (Mason, 1967; Melack, 1983, 1985; Lenz, 1984). The seasonal difference is caused primarily by changes in phytoplankton abundance. The depth of the euphotic zone (i.e., the depth to which 0.5 percent of incident photosynthetically active insolation reaches) ranges from 4 to 18 m (Jellison and Melack, in press). Measurements of underwater attenuation in the red, blue, and green spectral regions indicate greatest penetration in the green region (Mason, 1967).

The water in Mono Lake differs in physical properties from pure fresh water (Mason, 1967). First, the thermal capacity per gram is lower; hence fewer calories are required to warm Mono Lake than to warm an equivalent mass of pure water. Second, the viscosity of Mono Lake water is about 20 percent higher than that of pure water.

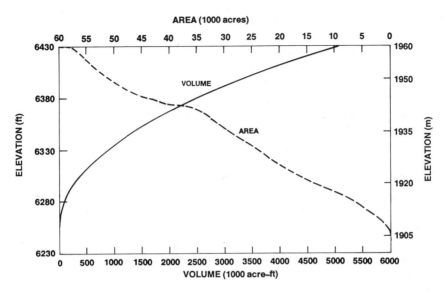

FIGURE 3.1 Area capacity for Mono Basin (Pelagos Corporation, 1987).

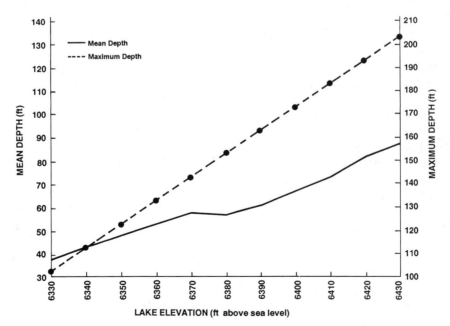

FIGURE 3.2 Morphometric parameters for Mono Basin (from data in Pelagos Corporation, 1987). Mean depth is on the left, and maximum depth is on the right.

Thus the sinking velocity of plankton may be slowed and the momentum transfer through the water column altered (Mason, 1967).

Density of water is a function of its temperature and the quantity of dissolved and particulate matter. Temporal changes in the variations in density with depth can cause ecologically important consequences. Thermal and chemical stratification occur in Mono Lake, and seasonal and inter-annual differences influence the biota.

Mono Lake was monomictic (circulated from top to bottom during one season each year) when studied in the early 1960s (Mason, 1967) and from 1978 to 1982 (Melack, 1983). It began to thermally stratify in late March or early April, remained stratified until November and was holomictic (mixed to the bottom throughout) during the winter. Maximum midsummer temperatures in near-surface, offshore water reach about 20°C. Minimum winter temperatures are near 0°C (Figure 3.3). Nearshore areas and occasionally much of the western bay can be ice covered as freshwater inflows freeze.

Exceptionally heavy snowfall and reduced diversions by the City of Los Angeles during 1982-1983 led to a large input of fresh water. The fresh water mixed only partially with the saline lake water. Salinities in the near-surface region declined, and a chemocline (vertical gradient in major solute concentration) developed between 12 and 16 m. Therefore, the lake did not mix to the bottom, and meromixis (incomplete vertical mixing) was initiated in 1983. The mixolimnion (upper mixed layer in a meromictic lake) has deepened each subsequent autumn, and the chemocline was between 16 and 18 m in mid-1985 (Figure 3.4). A slight temperature inversion occurs in early spring, with colder mixolimnetic water overlying warmer monimo-limnetic (region below the chemocline in a meromictic lake) water. Within the mixolimnion a thermal stratification and mixing cycle, similar to that previous to 1983, occurs. The thermocline develops above the chemocline and gradually descends to the depth of the chemocline by early autumn. However, large inflows of fresh water in the spring of 1986 have incompletely mixed and caused a secondary diffuse chemocline in the upper 15 m.

Calculations of the time required to erode the chemical stratification and permit complete mixing again are possible but difficult and require meteorological data that currently

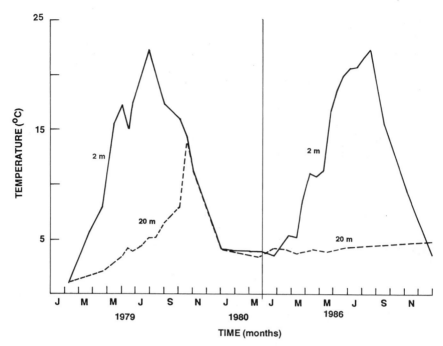

FIGURE 3.3 Temperature of Mono Lake at 2 and 20 m depth for 1979–1980 and 1986 at a station located approximately halfway between Paoha Island and the south shore.

are unavailable for Mono Lake. Furthermore, the density of Mono Lake is not well-known, and computations that estimate mixing as a function of energy inputs require accurate measurements of the density structure. Densities reported by Mason (1967) apply to the lake at a level of 6389.9 ft, and his data as well as those in Herbst (1986) and LADWP (1986) are expressed only to three decimal places. Only the measurements made with a vibrating densimeter (Picker et al., 1974) to a precision of $\pm 5 \times 10^{-6}$ g/cm^3 on water obtained when the lake stood at 6373.3 ft above sea level are of sufficient quality to calculate mixing rates (F. J. Millero, University of Miami, personal communication, 1982); values for additional levels are needed.

Motions of lake water such as horizontal currents, surface and internal waves, and turbulent eddies affect distributions of physical, chemical, and biological entities.

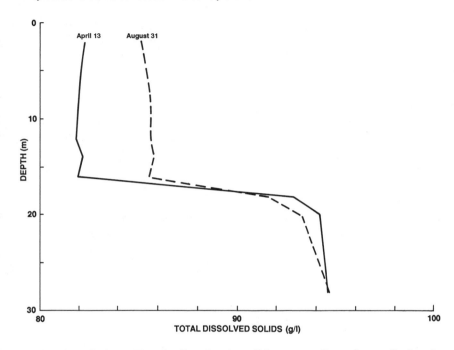

FIGURE 3.4   Total dissolved solids as a function of depth in Mono Lake in 1985.

Mason (1967) hypothesized that wind-driven currents and internal waves played a role in the erosion and deposition of sediments evident in the shape of the basin. Occurrences of features such as fronts as indicated by foam lines (Lenz, 1980) and large-scale differences in plankton abundance (Almanza and Melack, 1984; Lenz et al., 1986) indicate the presence of gyres or other circulation patterns. No direct measurements of currents are available.

Vertical mixing across the thermocline or chemocline is an important mechanism for injection of nutrients into the euphotic region where phytoplankton grow and for erosion of chemical or thermal stratification. The coefficient of eddy conductivity has been used to calculate vertical mixing rates in Mono Lake. Jellison and Melack (1986) applied the flux gradient method (Jassby and Powell, 1975) to temperature profiles obtained from 10 stations over 3 years. This technique is appropriate only during the period of net accumulation of heat by the lake. Results for the spring and early summer of 1983, 1984, 1985, and 1986 indicate a marked decrease in vertical mixing in the region near the chemocline once meromixis is established. When combined with vertical profiles of ammonium, this result produces a reduced supply of this limiting nutrient as a consequence of meromixis. Further discussion of the interactions among the thermocline, chemocline, ammonium, and algae in terms of nutrient supply to the phytoplankton is presented in chapter 4.

The motions and thermal structure of Mono Lake are closely related to the motions and thermal structure of the overlying atmosphere. Surface winds may hasten the overturning or mixing of the lake. For example, strong northerly winds of 1 or 2 days' duration following a deepening cyclone in fall have been identified with the irregular overturning of Lake Tahoe (Paerl et al., 1975).

CHEMICAL SYSTEM

Chemical Composition of the Water

The major components determining the chemistry of the water in Mono Lake are the major ions, major nutrients (nitrogen and phosphorus), trace elements, and dissolved oxygen. These components are discussed in the following sections.

Major Ions

Because the present-day chemistries of concentrated alkaline, saline lakes have been determined by the geochemical evolution of lake waters over geological time (10^3 to 10^4 years), Mono Lake water is not strongly affected by relatively recent changes (on the order of 10^0 to 10^2 years) in the chemistries of influent streams and hot springs. This has been demonstrated by Scholl et al. (1967), who calculated that the chloride age of the lake was about 31,000 years (Cl^- in the lake/the annual input of Cl^- to the lake), and Mason (1967), who made similar calculations. For practical purposes the major ion composition of Mono Lake is determined by dilution, evaporation, and the concomitant dissolution or precipitation of minerals.

In 1980 (LADWP, 1984) and presumably at present, the major ion composition of Mono Lake is approximately as follows: 96.8 percent Na^+, 3.0 percent K^+, and 0.2 percent Mg^{2+}; 46.3 percent CO_3^{2-} plus HCO_3^-, 37.8 percent Cl^-, and 15.9 percent SO_4^{2-} (as the equivalent percentages of the cations and anions, respectively). In simple terms, Mono Lake water is a triple water because it contains relatively large concentrations of carbonate plus bicarbonate, chloride, and sulfate. Triple waters possibly form in alkaline, saline lakes in which sulfate reduction has been low over much of geologic time.

Salinity as total dissolved solids (TDS) can be reported as grams per liter (mass/volume) or grams per kilogram (mass/mass). In this report, salinity data are given in grams per liter because experimental investigations involving the biota of Mono Lake have consistently used this unit. In general, presenting salinity data for saline waters in grams per kilogram is preferred. One can convert grams per liter to grams per kilogram by dividing by the density (liters per kilogram).

Nitrogen and Phosphorus

Nitrogen and phosphorus are often found to limit the abundance or productivity of algae. For Mono Lake, reliable values for inorganic nitrogen have only recently become available (Jellison and Melack, in press); previously published data (e.g., Winkler, 1977) were obtained with unsatisfactory methods. Concentrations of inorganic nitrogen are usually low in the upper mixed layer. Nitrate is less than 1 μM, and ammonium is less than 3 μM, except in the late summer, when values can reach 8 to 12 μM. In the anoxic water below the thermocline or chemocline, ammonium concentrations can be very high and have exceeded 200 μM in recent years. Phosphate concentrations are substantial (about 800 to 1000 μM) throughout the lake.

Studies of phytoplankton samples from Mono Lake experimentally enriched with ammonium show an increase in growth, evidence that phytoplankton production is limited by the amount of ammonium in the lake (Jellison and Melack, 1986). To evaluate the relative importance to the algae of nitrogen supply from external and internal sources, Jellison and Melack (1986) compared nitrogen data on rain and stream inflows to nitrogen data on brine shrimp excretion and vertical mixing. Inputs from the watershed and airshed were negligible. In the spring, adult brine shrimp are absent, and the supply of ammonium from the deeper water via mixing was less than algal demand. During midsummer, brine shrimp excretion was adequate to meet the algal needs, and upward vertical mixing was low. In the autumn, as the mixed layer deepens and brine shrimp numbers decline, the supply of ammonium from vertical mixing and entrainment became larger than the brine shrimp excretion.

Trace Elements

Alkaline, saline lakes generally contain comparatively large concentrations of a variety of minor elements. In Mono Lake the most abundant minor element is boron at approximately 460 mg/l (LADWP, 1984). Boron concentrations in Mono Lake are among the highest recorded for any lake (Whitehead and Feth, 1961). The alkaline earth metals are all found in lower abundance than boron (Sr^{2+} = 120 mg/l, Mg^{2+} = 33.4 mg/l, and Ca^{2+} = 4.1 mg/l), based on average values in Dana et al. (1977). The halogens are observed in moderate to high concentrations (F^- = 48 mg/l and Br^- = 40 mg/l). Fluoride concentrations of this magnitude should prevent some organisms from inhabiting the lake (Kilham and Hecky, 1973). Other nonnutrient minor elements listed in Dana et al. (1977) are arsenic (15.5 mg/l), lithium (10 mg/l), iodine (7 mg/l), and tungsten (4 mg/l).

Compilations of trace metal concentrations are presented in Mason (1967) and Dana et al. (1977). Average values for Mono Lake water (in micrograms per liter) are as follows: Fe = 420, Al = 40, Ti = 30, Mn = 20, and Cu = 10. How-

ever, these values should be redetermined using modern
analytical techniques.

Dissolved Oxygen

In hypersaline waters, such as those in Mono Lake, oxygen reaches saturation at concentrations well below those
in fresh water.   Hence, dissolved oxygen concentrations,
while near saturation, are moderate to low in the upper
mixed layer (Melack, 1983; Lenz, 1984).   Spring values are
between 4 and 7 mg/l; summer and autumn values are between 2 and 6 mg/l.   Below the thermocline or chemocline
the water is without oxygen.

Inputs of Chemical Constituents from Surface Water
and Groundwater

The geochemical origin of the mineral content of mountain spring and stream waters flowing from the Sierra
Nevada is well known (Feth et al., 1964; Garrels and
MacKenzie, 1967; Stoddard, in press).   Snowmelt derived
from precipitation and groundwater become charged with
carbon dioxide and interact with the soil and primarily
granitic bedrock.   Chemical weathering ensues, and plagio-
clase, biotite, and K-feldspars, for example, may be weathered to kaolinite.   The mean chemical composition of the
perennial springs in the Sierra Nevada as determined by
Feth et al. (1964) is very similar to the chemical composition of the major streams that flow from the mountains
into Mono Lake.   The order of concentration of cations
and anions in moles is: $Ca^{2+} > Na^+ > Mg^{2+} > K^+$ and
$HCO_3^- > SO_4^{2-} > Cl^-$.
Analyses of surface water and groundwater are presented in LADWP (1984).   The observed chemical differences between surface and well waters indicate that calcium-
dominated surface waters evolve into sodium-dominated well
waters owing to evaporative concentration and subsequent
precipitation of calcium carbonate.   Calcium-dominated surface waters have conductivities (an index of concentration)
of less than 300 μmhos/cm (13 out of 13 cases), while

sodium-dominated well waters have higher conductivities (17 out of 18 cases). Four out of forty cases cannot be explained easily. Sodium was the dominant cation in one surface water and one well water with low conductivities (<300 μmhos/cm), and calcium was the dominant cation in two well waters with high conductivities.

The chemical composition of the springs varies some-what, but the springs are primarily dominated by sodium and carbonate plus bicarbonate (32 out of 35). Paoha 1, Solo TT, and Dry Creek are the only spring waters in which chloride predominates over carbonate plus bicarbonate (Lee, 1969; LADWP, 1984). The temperatures of the springs range from 7.8°C at Bridgeport (LADWP, 1984) to 86°C at the hot springs on Paoha Island (Lee, 1969). The seasonal variability of hot spring temperatures and chemistries has not been determined.

Although the major ion chemistries of many springs are known, the contribution of the springs to the overall chemical budget of the lake cannot be calculated. This calculation would require data on the contribution of the springs to the basin's moisture budget, information that is not available (see chapter 2).

Data for the chemistry of atmospheric precipitation in the vicinity of Mono Lake are given in Melack et al. (1982).

Tufa Pinnacles Associated with Springs

Tufa towers are formed by precipitation of calcite, high-magnesium calcite, and aragonite. This precipitation results from two related processes. First, there is the strictly inorganic precipitation of carbonate minerals that occurs when calcium-containing spring or stream waters mix with the carbonate-rich waters of Mono Lake. Calcium carbonate should precipitate directly when the ion activity product of the mixture of waters exceeds the solubility product of any of the carbonate minerals in question. However, in natural waters, supersaturation is commonly observed. High concentrations of phosphorus, for example, may inhibit the formation of calcium carbonate. The concentrations of phosphorus in Mono Lake are high (see above section on nitrogen and phosphorus). Second, the precipitation of calcium carbonate is enhanced by the activities of photosynthetic organisms. The formation of tufa pinnacles about the orifices of sublacustrine springs is a result of inorganic precipitation and the activities of mat-forming benthic algae, which can guide the precipitation of calcium carbonate minerals and thus determine the morphology of the resulting tufa pinnacles. Because photosynthesizing algae take up CO_2, they lower CO_2 tensions in the microenvironment of calcium carbonate precipitation. The lowered CO_2 tensions result in elevated CO_3^{2-} activities and enhanced precipitation of carbonate minerals. The mat-forming algae in Mono Lake are primarily filamentous green algae, diatoms, and blue-green algae (Scholl and Taft, 1964; Herbst, 1986).

Losses of Salts and Other Substances

Substances are lost from the lake through sedimentation, deflation, and aerosol production. The amount of organic and inorganic materials lost from the lake water to the sediments of Mono Lake remains unknown. Comprehensive sedimentological and paleolimnological investigations have not been undertaken within the main basin of the lake, although uplifted Pleistocene sediments on Paoha Island (Lajoie, 1968; Reed, 1977) have been studied. Reed (1977, citing various authors) gives an average sedimentation rate for Mono Lake of approximately 0.40 m per thousand years. On the basis of his own research on the organic biogeochemistry of Mono Lake sediments, Reed estimates that 4 percent of the organic matter produced as a result of primary production becomes sediment.

Losses of salts by deflation and aerosol production occur in all closed basin lakes, but few quantitative data are available. Langbein (1961) discusses the primary mechanisms responsible for the loss of salts to the atmosphere from concentrated lakes. He also points out that salts precipitated at the margins of lake basins may become unavailable to a lake in two ways. First, they may be blown away by the wind (deflation). Second, they may become covered over in such a way that they do not redissolve when the lake level rises. The consequences of deflation at Mono Lake are clearly illustrated by the air quality problems that occur in the vicinity of the lake (Kusko and Cahill, 1984).

Geochemical Evolution

The geochemical evolution of waters in the Mono Basin to a brine of Mono Lake's present-day chemical composition can be explained in terms of the rock weathering / evaporative concentration / mineral precipitation model of lake water evolution (Garrels and MacKenzie, 1967; Eugster and Hardie, 1978). However, the geochemical evolution of the water in Mono Lake is complicated by the presence of several volcanic hot springs that flow directly into the lake. In the first part of the following discussion the geochemical evolution of waters in the Mono Basin will be examined without taking the possible effects of the hot springs into account. At the end of this section the possible effects of the hot springs chemistry on the lake will be explored.

Atmospheric precipitation falling on the drainage basin interacts primarily with igneous and metamorphic silicate rocks according to the following generalized equation:

$$\text{cation Al silicate (silicate rocks)} + H_2CO_3^* + H_2O \rightarrow$$
$$HCO_3^- + H_4SiO_4 + \text{cation} + \text{Al silicate (clay minerals)}$$

where $H_2CO_3^*$ represents CO_2 (aq) + H_2CO_3 concentrations above equilibrium levels (Stumm and Morgan, 1981). Garrels and MacKenzie's (1967) classic paper on the origin of the chemical compositions of spring and lake waters models the geochemical evolution of waters flowing from the Sierra Nevada. Judging from their chemical compositions (see LADWP, 1984), waters in the Mono Basin (including those flowing from the Mono Craters) evolved in a similar manner.

Available chemical analyses (Lee, 1969; LADWP, 1984) of the surface waters, wells, and springs are important because they indicate that the dilute waters of the Mono Basin in general contain higher molar concentrations of bicarbonate than calcium. As a result of calcite (calcium carbonate) precipitation (see Eugster and Hardie, 1978), essentially all of the calcium initially present will be removed from solution as these waters are concentrated by evaporation. The shift from calcium dominance in the stream waters to sodium dominance in the more concentrated wells and springs is presumably a consequence of the precipitation of calcite and related minerals.

The chemical composition and geochemical evolution of Mono Lake is controlled primarily by mineral precipitation. Sodium, potassium, sulfate, carbonate plus bicarbonate, and chloride are concentrated by evaporation, while calcium, magnesium, and silicon are not. If Mono Lake becomes 1.4 times more concentrated than it is at present (about 125 g/l), minerals containing sodium will begin to precipitate. R. J. Spencer (University of Calgary, personal communication, 1986) has constructed a geochemical model that indicates that trona ($NaHCO_3 \cdot Na_2CO_3 \cdot 2H_2O$), mirabilite ($Na_2SO_4 \cdot 10H_2O$), and natron ($Na_2CO_3 \cdot 10H_2O$) start to precipitate at low temperatures (near 0°C) at these salinities. The solubilities of these minerals depend mainly on ion activities, temperature, and CO_2 tension. Even though potassium is concentrated by evaporation, it is continuously lost from inflowing waters and lake brine as a result of exchange and fixation on clays (Spencer et al., 1985). Sulfate can be removed as a result of sulfate reduction and subsequent precipitation of metal sulfides or the loss of hydrogen sulfide to the atmosphere, but the high sulfate concentrations in the lake indicate that sulfate reduction is not a particularly important process in Mono Lake. Mirabilite precipitation, on the other hand, can potentially remove considerable quantities of sulfate. Carbonate is currently lost from solution by the precipitation of carbonate minerals (see below). Once the concentration of the lake reaches approximately 125 g/l, carbonate plus bicarbonate will begin to precipitate in winter as trona and natron. Chloride is very soluble in Mono Lake water, and it does not precipitate until saturation with respect to halite (NaCl) is reached. This will not occur until the lake is about 10 times more concentrated than it is at present.

Calcium, magnesium, and silicon are not concentrated by evaporation, and their concentration in Mono Lake is determined by a variety of processes. Calcium precipitates as calcite, high-magnesium calcite, and aragonite. Some magnesium is removed as aragonite, but much of it may be lost from solution owing to the formation of clay minerals rich in magnesium and silicon (see Jones and Wier, 1983). Silicon is taken up by diatoms and then lost to the sediments, or it is utilized in the formation of clay minerals.

In addition to mineral precipitation and dissolution, ions can be lost or gained by lakes owing to interactions with pore fluids. In lakes with fluctuating level, ions will diffuse into or out of the sediments as the concentration of the lake changes (Spencer et al., 1985).

Volcanic hot springs are very complex, but they can be viewed as being composed primarily of meteoric waters (derived from the atmosphere) that circulate to considerable depths in the earth, where they interact with magmatic gases and are heated (White, 1957a,b). The mineral salts in hot-spring waters come largely from the original meteoric waters (e.g., Mono Lake water), condensed magmatic gases, and rock-water interactions at depth (White, 1957a; Ellis and Mahon, 1964). Neither the moisture budget of the Mono Basin nor the geochemistry of the hot springs is known sufficiently to determine if the mineral salts from the hot springs (in excess of those contained in the original meteoric waters) have markedly affected the chemistry of Mono Lake water.

The major effects that volcanic hot springs of the sodium chloride type should have on the waters of Mono Lake are to increase the relative proportions of sodium and chloride among the major cations and anions, respectively, and to increase the loading of silicon, boron, fluoride, bromide, iodide, lithium, and arsenic to the lake (White, 1957a,b). Hot-spring waters such as those flowing from Paoha No. 1 on June 21, 1968 (Lee, 1969) are enriched in chloride ions (equivalent percentage among the anions equal to 47 percent) in comparison to Mono Lake water (equivalent percentage among the anions equal to 38 percent). Paoha No. 1 also contains high concentrations of fluoride (26.5 mg/l) and boron (227 mg/l), but these high concentrations are expected if Mono Lake was the original source of much of the meteoric water flowing from the hot spring. The minor elements (listed above) in Mono Lake water can come either from the volcanic hot springs or from rock weathering in the Mono Basin. Sufficient hydrological and geochemical information is not available to distinguish one source from the other.

REFERENCES

Almanza, E., and J. M. Melack. 1985. Chlorophyll differences in Mono Lake (California) observable on Landsat imagery. Hydrobiologia 122:13-17.

Dana, G. L., D. B. Herbst, C. Lovejoy, B. Loeffler, and K. Otsuki. 1977. Physical and chemical limnology. Pp. 40-42 in An Ecological Study of Mono Lake, California, D. W. Winkler, ed. Institute of Ecology Publication No. 12. Davis, Calif.: University of California, Institute of Ecology.

Ellis, A. J., and W. A. J. Mahon. 1964. Natural hydrothermal systems and experimental hot-water/rock interactions. Geochim. Cosmochim. Acta 28:1323-1357.

Eugster, H. P., and L. A. Hardie. 1978. Saline lakes. Pp. 237-293 in Lakes: Chemistry, Geology, Physics, A. Lerman, ed. New York: Springer-Verlag.

Feth, J. H., C. E. Roberson, and W. L. Polzer. 1964. Sources of mineral constituents in water from granitic rocks, Sierra Nevada, California and Nevada. Geochemistry of water. U.S. Geological Survey Water-Supply Paper 1535-I. Washington, D.C.: U.S. Geological Survey. 70 pp.

Garrels, R. M., and F. T. MacKenzie. 1967. Origin of the chemical compositions of some springs and lakes. Pp. 222-242 in Equilibrium Concepts in Natural Water Systems. Advances in Chemistry Series No. 67. Washington, D.C.: American Chemical Society.

Herbst, D. B. 1986. Comparative Studies of the Population Ecology and Life History Patterns of an Alkaline Salt Lake Insect: *Ephydra (Hydropyrus) Hians* Say (Diptera: Ephydridae). Ph.D. dissertation, Oregon State University, Corvallis. 222 pp.

Jassby, A., and T. Powell. 1975. Vertical patterns of eddy diffusion during stratification in Castle Lake, California. Limnol. Oceanogr. 20:530-543.

Jellison, R., and J. M. Melack. 1986. Nitrogen supply and primary production in hypersaline Mono Lake. Eos Trans. Am. Geophys. Union 67:974.

Jellison, R., and J. M. Melack. In press. Photosynthetic activity of phytoplankton and its relation to environmental factors in hypersaline Mono Lake, California. In Saline Lakes, J. M. Melack, ed. Developments in Hydrobiology. Dordrecht, Netherlands: Dr W. Junk Publishers.

Jones, B. F., and A. H. Weir. 1983. Clay minerals of Lake Abert, an alkaline, saline lake. Clays Clay Miner. 31:161-172.

Kilham, P., and R. E. Hecky. 1973. Fluoride: geochemical and ecological significance in East African waters and sediments. Limnol. Oceanogr. 18:932-945.

Kusko, B. H., and T. A. Cahill. 1984. Study of Particle Episodes at Mono Lake. Final Report to California Air Resources Board on Contract A1-144-32. Davis, Calif.: University of California, Air Quality Group, Crocker Nuclear Laboratory.

Lajoie, K. R. 1968. Late Quaternary Stratigraphy and Geologic History of Mono Basin, Eastern California. Ph.D. dissertation, University of California, Berkeley. 379 pp.

Langbein, W. B. 1961. Salinity and Hydrology of Closed Lakes. U.S. Geological Survey Professional Paper 412. Washington, D.C.: U.S. Government Printing Office. 20 pp.

Lee, K. 1969. Infrared Exploration for Shoreline Springs: A Contribution to the Hydrogeology of Mono Basin, California. Ph.D. dissertation, Stanford University. 216 pp.

Lenz, P. H. 1980. Ecology of an alkali-adapted variety of *Artemia* from Mono Lake, California, USA. Pp. 79-96 in The Brine Shrimp *Artemia*. Vol. 3. Ecology, Culturing, Use in Aquaculture, G. Persoone, P. Sorgeloos, O. Roels, and E. Jaspers, eds. Wetteren, Belgium: Universa Press.

Lenz, P. H. 1984. Life-history analysis of an *Artemia* population in a changing environment. J. Plankton Res. 6:967-983.

Lenz, P. H., S. D. Cooper, J. M. Melack, and D. W. Winkler. 1986. Spatial and temporal distribution patterns of three trophic levels in a saline lake. J. Plankton Res. 8:1051-1064.

Los Angeles Department of Water and Power. 1984. Background Report on Mono Basin Geology and Hydrology. Report of the Aqueduct Division, Hydrology Section. Los Angeles, Calif.

Los Angeles Department of Water and Power. 1986. Report on Mono Lake Salinity. Los Angeles, Calif.

Mason, D. T. 1967. Limnology of Mono Lake, California. University of California Publications in Zoology No. 83. Berkeley, Calif.: University of California Press.

Melack, J. M. 1983. Large, deep salt lakes: a comparative limnological analysis. Hydrobiologia 105:223-230.

Melack, J. M. 1985. The Ecology of Mono Lake, California. Pp. 461-470 in National Geographic Society Research Reports, Vol. 20. 1979 Projects. Washington, D.C.: National Geographic Society.

Melack, J. M., J. L. Stoddard, and D. R. Dawson. 1982. Acid precipitation and buffer capacity of lakes in the Sierra Nevada, California. Pp. 465-471 in Proceedings of the International Symposium on Hydrometeorology, Denver, Colo., June 13-17, 1982, A. I. Johnson and R. A. Clarke, eds. Bethesda, Md.: American Water Resources Association.

Paerl, H. W., R. C. Richards, R. L. Leonard, and C. R. Goldman. 1975. Seasonal nitrate cycling as evidence for complete vertical mixing in Lake Tahoe, California-Nevada. Limnol. Oceanogr. 20:1-8.

Pelagos Corporation. 1987. A Bathymetric and Geologic Survey at Mono Lake, California. Report prepared for Los Angeles Deparment of Water and Power. San Diego, Calif.

Picker, P., E. Tremblay, and C. Jolicoeur. 1974. A high-precision digital readout flow densimeter for liquids. J. Solution Chem. 3:377-384.

Reed, W. E. 1977. Biogeochemistry of Mono Lake, California. Geochim. Cosmochim. Acta 41:1231-1245.

Scholl, D. W., and W. H. Taft. 1964. Algae, contributors to the formation of calcareous tufa, Mono Lake, California. J. Sediment. Petrol. 34:309-319.

Scholl, D. W., R. Von Huene, P. Saint-Amand, and J. B. Ridlon. 1967. Age and origin of topography beneath

Mono Lake, a remnant Pleistocene lake, California. Geol. Soc. Am. Bull. 78:583-600.

Spencer, R. J., H. P. Eugster, B. F. Jones, and S. L. Rettig. 1985. Geochemistry of Great Salt Lake, Utah. I. Hydrochemistry since 1850. Geochim. Coschim. Acta 49:727-737.

Stoddard, J. L. In press. Alkalinity dynamics in an un-acidified alpine lake, Sierra Nevada, California. Limnol. Oceanogr.

Stumm, W., and J. J. Morgan. 1981. Aquatic Chemistry: An Introduction Emphasizing Chemical Equilibria in Natural Waters, 2nd ed. New York: Wiley. 780 pp.

White, D. E. 1957a. Thermal waters of volcanic origin. Geol. Soc. Am. Bull. 68:1637-1658.

White, D. E. 1957b. Magmatic, connate, and metamorphic waters. Geol. Soc. Am. Bull. 68:1659-1682.

Whitehead, H. C., and J. H. Feth. 1961. Recent chemical analyses of waters from several closed-basin lakes and their tributaries in the western United States. Geol. Soc. Am. Bull. 72:1421-1425.

Winkler, D., ed. 1977. An Ecological Study of Mono Lake, California. Institute of Ecology Publication No. 12. Davis, Calif.: University of California, Institute of Ecology.

4

Biological System of Mono Lake

INTRODUCTION

Mono Lake is a productive aquatic ecosystem but with very few species. The lake has two major habitats--an open water pelagic region and a nearshore littoral region. Trophic structure, the linkages within the food web, is different in these two habitats.

In the pelagic waters, phytoplankton are the primary producers, using sunlight to reduce inorganic carbon to organic matter. These algae are grazed by the brine shrimp, *Artemia monica*, which are preyed upon mainly by eared grebes (*Podiceps nigricollis*) and California gulls (*Larus californicus*). No fish live in Mono Lake. The current combination of high salinity and alkalinity makes it impossible for fish to survive. Inputs of organic material to the profundal sediments, those sediments under the pelagic zone, consist largely of fecal pellets and cysts of brine shrimp and detritus. No zoobenthos has been recorded in the profundal sediments, which are anoxic much or all of the year. The role of protozoans and bacteria as food for brine shrimp or as decomposers of organic matter remains undetermined, but is likely to be of importance.

In the littoral region, the overlying waters have the same planktonic organisms as the pelagic zone and an intermittent complement of organisms associated with the bottom. The benthic habitat is highly variable as a function of depth and substrate (Herbst, 1986; Pelagos Corporation, 1987). The principal constituents are a microbial

69

community, an algal flora, and a brine fly, *Ephydra hians*, which feeds upon the benthic algae and probably bacteria and detritus derived from a number of sources, some likely to be terrestrial.   The brine fly is prey to a variety of birds including phalaropes, and to a lesser extent eared grebes and gulls.

Although the trophic structure of Mono Lake is simple in comparison with that in many aquatic ecosystems, the lack of sufficient information on key components such as bacteria and protozoans precludes the formulation of a complete, quantitative description of carbon or nitrogen flow through the whole food web.   Two trophic links that have received some quantitative attention are the algae-brine shrimp and brine shrimp-bird links.   Grazing by brine shrimp contributes to a decline in phytoplankton during the spring and maintains a low algal abundance during the summer (Lenz, 1982; Jellison, 1985).   The regeneration of ammonium by the brine shrimp, in turn, sustains the growth of the phytoplankton (Jellison and Melack, 1986). The decline in brine shrimp in the autumn can be, in part, attributed to predation by the grebes (Cooper et al., 1984).

The remainder of this chapter discusses the ecological and physiological aspects of the components of the food web--and primary producers and decomposers (bacteria, phytoplankton, and phytobenthos), primary consumers (brine shrimp and brine fly), and secondary consumers (aquatic bird populations).

ECOLOGICAL ASPECTS OF AQUATIC PELAGIC AND LITTORAL ORGANISMS

Primary Producers and Decomposers

Bacteria

The abundance and significance of bacteria in alkaline, saline lakes are not well-known.   Bacteria probably function as both decomposers and primary producers in the food web of Mono Lake.

Recent research by R. S. Oremland and his associates indicates that the same major processes that are carried out by anaerobic bacteria in fresh water and marine habitats also occur in alkaline, saline lakes. They have examined, for example, methanogenesis, sulfate reduction, and other anaerobic processes in Big Soda Lake, Nevada (Oremland et al., 1982, 1985; Iversen et al., in press). In Mono Lake, R. S. Oremland (U.S. Geological Survey, Menlo Park, personal communication) has discovered that large quantities of methane are leaving the sediments even though relatively small amounts of methane are produced at the sediment-water interface. He argues that most of the methane-rich gas seeps in the lake produce biogenic methane that is derived from the anaerobic decomposition of fossil organic matter by bacteria. However, the methane from one seep associated with a hot spring had a more thermogenic character, indicating a chemical process that does not involve bacteria. The presence of numerous gas seeps on the floor of Mono Lake is supported by the discovery that large areas of bottom sediments are disturbed by gas bubbles (Pelagos Corporation, 1987).

Pelagic, aerobic bacteria are often abundant in alkaline, saline lakes. In freshwater lakes and in the ocean, concentrations of pelagic bacteria between 10^5 and 10^6 bacteria/ml are commonly observed. However, alkaline, saline lakes in east Africa contain from 10^7 to 10^8 bacteria/ml (Kilham, 1981). These large concentrations presumably represent a balance between the availability of organic substrates in these highly productive lakes and the abundance of heterotrophic organisms that consume bacteria (e.g., ciliates). Pyramid Lake in Nevada is the only alkaline, saline lake in the Great Basin in which a detailed study of bacteria has been carried out. Hamilton-Galat and Galat (1983) found from 5.1×10^5 to 2.5×10^7 bacteria/ml in Pyramid Lake. Bacterial numbers more or less tracked periods of algal production. One reason that bacterial numbers are not higher in Pyramid Lake is that the lake is only moderately productive (i.e., mesotrophic). For Mono Lake, R. S. Oremland (personal communication) and R. W. Harvey (U.S. Geological Survey, Menlo Park, personal communication) have observed bacterial concentrations of between 1.4 and 2.0×10^7 bacteria/ml. On average, these concentrations are considerably higher than most found in

Pyramid Lake and generally similar to those observed in the lakes in east Africa.

Phytoplankton and Phytobenthos

The algal community of Mono Lake includes few species, as is typical of hypersaline waters. The phytoplankton is dominated by a coccoid green alga, *Nannochloris* sp., cyanobacteria, and diatoms (Mason, 1967; Lovejoy and Dana, 1977; Melack, 1983). The benthic algae are composed of *Nitzschia frustulum*, other less common diatoms, filamentous cyanobacteria, and the green alga, *Ctenocladus circinnatus* (Herbst, 1986).

The seasonal dynamics of the phytoplankton in Mono Lake are unusual (Mason, 1967; Lovejoy and Dana, 1977; Melack, 1983, 1985; Jellison and Melack, in press) (Figure 4.1). During the winter, the phytoplankton are abundant throughout the lake, and after the onset of the seasonal thermocline in early spring, the algae increase in the upper water. This increase was reduced during 1984, 1985, and 1986 after the initiation of meromixis. As described in chapter 3, the chemical stratification reduced vertical mixing, which reduced the supply of the limiting nutrient, nitrogen, to the euphotic zone. A rapid decrease in algal abundance occurs in late May and June above the thermocline. During the summer, the phytoplankton are sparse in the upper waters and abundant in the deeper, cold and dim or dark waters. In midsummer, higher chlorophyll concentrations occur in a layer coinciding with the chemocline. In autumn, algal concentrations increase in the upper waters as thermal stratification weakens and brine shrimp numbers decline (Figure 4.1).

Primary productivity measurements spanning the period from 1983 to 1985 vary from 340 to 540 g carbon/m^2/yr (Jellison and Melack, in press). Mono Lake would thus be classified as eutrophic. Production was higher during the spring of 1983 than in 1984 and 1985; the difference may be at least partially attributed to meromixis.

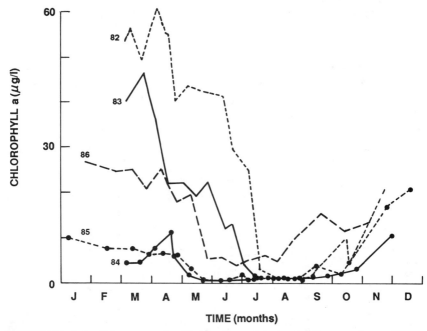

FIGURE 4.1 Mean mixolimnetic chlorophyll *a* for 1982, 1983, 1984, 1985, and 1986.

Primary Consumers

Zooplankton

The Mono Lake brine shrimp, *Artemia monica*, is the major zooplankton species (Mason, 1967; Lenz, 1980, 1982). *A. monica*, a member of the *A. franciscana* superspecies, is now considered a sibling species (Bowen et al., 1985). The zooplankton also includes protozoans and has included rotifers (Mason, 1967).

The abundance of brine shrimp in Mono Lake varies seasonally (Lenz, 1982, 1984; Figure 4.2). The brine shrimp hatch from overwintering cysts from January through May. By mid-May, the first adult brine shrimp are present. For

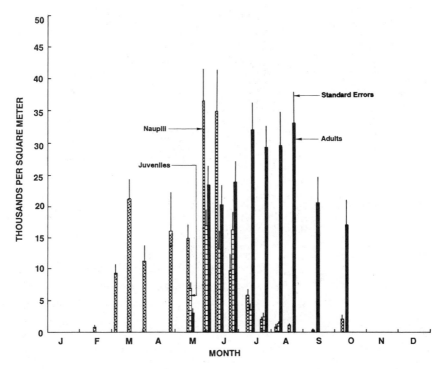

FIGURE 4.2 Seasonal abundance of brine shrimp at Mono Lake in 1985. Lakewide mean of 10 stations (three vertical net tows per station).

approximately one month females bear live young, which mature rapidly in the warm upper mixed layer. In June females switch to oviparous reproduction. The diapause eggs lie dormant on the bottom of the lake until the following winter. During the summer, brine shrimp are abundant in the oxygenated upper waters and very sparse or absent in the anoxic deeper waters. By September, the brine shrimp begin to decline in numbers and are almost absent from the plankton by December.

Studies conducted with similar methods since 1978 permit interannual comparisons of brine shrimp abundances and reproductive characteristics (Lenz, 1984; G. Dana, R. Jellison, and J. M. Melack, University of California, Santa Barbara, unpublished). Statistically significant interannual differences in abundances of first-generation adults (late May to June populations) occurred. In 1979, 1984, and 1986 brine shrimp numbered between 19,000 and 31,000 animals/m^2, whereas from 1980 to 1983 numbers were only 2,400 to 5,700 animals/m^2. Maximum midsummer abundances (first- and second-generation adults) were much higher in 1981 and 1982 than in other years. A number of factors are associated with these variations. First-generation adult abundances depend on the number of cysts available for hatching, hatching success, and survival to adulthood. In laboratory experiments, Dana and Lenz (1986) determined that salinities in the period from 1979 to 1986 are not indicated as a cause for changes in hatching success. Emergence trap trials in spring 1985 showed that very low hatching occurred in sediments in anoxic water below the chemocline (Dana et al., in press). In contrast, large numbers of cysts lying in sediments under the oxygenated mixolimnion hatched. The number of cysts available depends on the production of cysts during the previous year and possibly past years and on the viability of the cysts. Cyst production is related to brood size, numbers of ovigerous females, the percentage of those females producing cysts, and the time interval between broods. Brood size varied from 30 to 140 eggs per brood from 1983 to 1986 and is explained primarily by differences in female length and algal abundance. Second-generation abundances depend on the abundance, percent ovoviviparity, and fecundity of the first-generation females. Recruitment to adults depends on survival of naupliar and juvenile stages. Differences in all these factors occurred from 1982 to 1986. The switch from ovoviviparity (live bearing) to oviparity (cyst production) occurred at the time of decreasing phytoplankton in all years studied. In years with a substantial spring hatch, the first generation dominates the population. When the spring hatch is relatively low, first-generation adults are less abundant, algal densities remain higher later into the spring, and a large second generation can occur.

The spatial distribution of brine shrimp is heterogeneous on large (square kilometers) and small (square meters) spatial scales and varies on time scales from hours to days to seasons (Lenz, 1982; Melack, 1985; Lenz et al., 1986; Conte

et al., in press).   These differences in concentrations of
brine shrimp result in variable profitability of foraging for
the birds (see section on bird populations below).   Small-
scale patches are associated with sublacustrine springs
where upwelling varies widely in strength and hence in the
entrainment of brine shrimp.   Mason (1966) hypothesized
that the very dense plumes that formed near shore, but not
in association with springs, result from thermal currents
and behavioral responses of the animals.   Foam lines contain concentrations of living and dead brine shrimp as well
as other debris and can stretch for hundreds of meters.
These features seem to delimit water masses.   Large-scale
patchiness has been documented by sampling transects and
lakewide grids.   Abundance differences between the eastern
and western halves of the lake are common.   The degree of
variability differs seasonally and appears greater during
transition periods such as spring and autumn.

Current sampling programs are designed to assess the
lakewide abundance of brine shrimp and include biweekly or
monthly samples from 10 pelagic stations.   Regular sampling
is not performed in water overlying the littoral region or
at sites of aggregation such as springs.   Therefore, while
providing statistically sound estimates of the overall abundance of brine shrimp, the sampling does not include sites
that may be of particular importance to some birds some of
the time.   No efforts are in progress to sample zooplankton
other than brine shrimp.

Zoobenthos

The benthic community of Mono Lake includes several
species of dipteran insects, as is typical of hypersaline
waters.   The predominant dipteran is the brine fly, *Ephydra
hians*, but other species are present, such as the deer fly
(*Chrysops* sp.) and the long-legged fly (*Hydrophorus plum-
beus*).   The biting midge (*Culicoides occidentalis*) is also
found among the macroinvertebrates (Herbst, 1986).

The seasonal dynamics of the macroinvertebrates are not
well-known.   However, recent research on brine flies by
Herbst (1986) using the third instar and pupae as population indices showed a phase of rapid population growth

occurring in the spring (May and June), a summer maximum (July through September), a gradual decline in the autumn, and minimal abundance from late winter through early spring. Since seasonal dynamics of the phytobenthic bacterial and algal populations (diatoms and filamentous algae) are unknown, one cannot determine if the zoobenthic community, as reflected by numbers of brine flies, is tracking periods of algal production.

The spatial distribution of brine flies is heterogeneous on large (square kilometers) and small (square meters) scales (Herbst, in press). Small-scale patches are associated with tufa pinnacles and nearshore grasses, which are excellent substrates for larval and pupal attachment. Large-scale patchiness has recently been documented by video and lakewide bathymetric transects (Pelagos Corporation, 1987). Large mats of pupae have been found on dead submerged grasses along the eastern shore and on underwater tufa and hard-surface sediments at depths sometimes greater than 10 m in the central and eastern provinces, as shown in Figure 4.3. Abundance differences observed between tufa-hard rock shoal regions and soft mud-sand lake bottom sediments common to the eastern and central provinces are probably due to the larvae's inability to attach to smooth surfaces. The placement of eggs by brine fly females at depths greater than a few meters has been observed (Pelagos Corporation, 1987). This observation raises questions about the typical mechanism of oviposition reported for other ephydran flies, including whether females utilize respiratory mechanisms other than gas bubble entrapment for vertical descent. If the lake level dropped, the loss of hard-surface sediments would reduce brine fly habitat.

PHYSIOLOGICAL ASPECTS AND SALINITY TOLERANCES OF AQUATIC PELAGIC AND LITTORAL ORGANISMS

Primary Producers

Two kinds of evidence are available to evaluate the effects of increased salinity on phytoplankton: (1) algal responses to experimental increases in the salinity of Mono

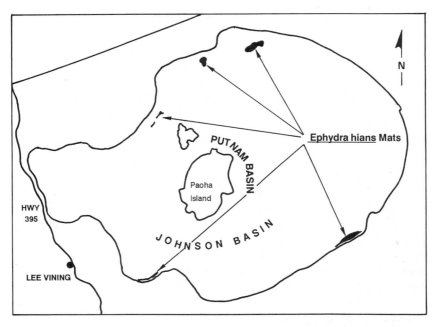

FIGURE 4.3 Locations of mats of brine fly (Pelagos Corporation, 1987).

Lake water and (2) salinity tolerances of phytoplankton known to occur in saline waters. Melack (1985) and Melack et al. (1982) studied microcosms containing Mono Lake water and microflora, some with brine shrimp and some without. These experiments, initiated in March 1981, gradually increased the salinity to 1.5 times the lake's salinity at the time (97 g/l) over a 3-month period. The results indicate a 10 percent decrease in gross primary productivity of microcosms with each 10 percent increase in salinity over the range of approximately 97 to 140 g/l TDS. Chapman (1982) reported results from laboratory experiments with isolates of the dominant algae--a green alga and a diatom--in synthetic and concentrated Mono Lake water. At 150 g/l, diatom growth was 95 percent of that at 97 g/l. Diatoms did not survive at a salinity of 185 g/l. At 150 g/l, the green alga growth was 74 percent of that at 97 g/l. The green alga continued to show growth beyond 185 to at least 237 g/l. Total salinity, arsenic, and fluoride contribute to these reductions in growth. Overall, Chapman's results indicate that a salinity of about 175 g/l results in fairly large decreases in growth of the two currently dominant algae.

A number of species of phytoplankton, not currently extant or dominant in Mono Lake, are known to grow well at salinities reaching 200 g/l (Hammer, 1986; Melack, in press). In particular, a green alga, *Dunaliella* sp., now occurs in low numbers in Mono Lake, and a related species, *D. parva*, grew best in Mono Lake concentrated to 150 g/l and still grew slowly at 235 g/l (Chapman, 1982).

Herbst (1986) isolated a clone of *Ctenocladus circinnatus* from Mono Lake and determined its growth rate and yield at salinities of 25, 50, 75, 100, and 150 g/l. The solutions were obtained by dilution or low-temperature evaporation of Mono Lake water. The experiments indicated decreased growth and yield at 75 and 100 g/l and no growth at 150 g/l.

Primary Consumers

A fundamental requirement for aquatic organisms living in saline lakes is to have sufficient "free water," or water that is available to sustain vital cellular activities. (In aquatic organisms, water molecules forming hydration shells are termed "bound water"; water molecules not associated with these shells are termed "free water.") This biological axiom is most evident in the osmotic effects of salinity upon the growth and larval development of brine shrimp and brine flies. If the salinity of the medium in which the organisms reside is sufficiently high, the thermodynamic forces responsible for the osmotic gradient will not allow sufficient free water to remain inside the organism. The loss of free water will in turn cause an inhibition or cessation of metabolic processes. This physical relationship sets an absolute upper limit on the salinity of salt lakes in which a self-sustaining population of halophilic organisms such as brine shrimp and brine flies can persist.

Physiological solutions within and surrounding cells are primarily dilute aqueous solutions. These solutions contain a large amount of free water and behave in a manner similar to pure water (P_0), which has predictable physical and chemical properties. Environmental salinities can cause organisms to alter their physiological solutions such that many of the internal fluids tend to become concentrated and the available free water diminishes. This is caused by free water molecules becoming associated with cellular solutes.

Measurements of free water in both biotic and abiotic saline solutions have been made by the determination of water activity (a_w) as reflected in the vapor pressure of saturated solutions relative to that of pure water (P/P_0). Table 4.1 consists of values for a_w, expressed as water vapor sorption ratios for various saturated salt solutions (Rockland, 1960; Winston and Bates, 1960). Additional values for solutions below saturation, especially sodium chloride, have been taken from studies by Lang (1967) and Clegg (1976a).

The amount of free water needed to sustain vital cellular and subcellular activities in *A. salina* has been established by Clegg (1974, 1976a,b, 1978). Upon analyzing the internal water content of various embryos, Clegg demonstrated that embryos required 0.65 g H_2O/g dry weight to be viable (Table 4.2). Activation of dormant embryos (cyst stage) required that environmental salinities have water activities (a_w) of 0.95. If water activities were lower, for instance below 0.93, only partial developmental activities were restored. The embryo could achieve only a degree of hydration that was equivalent to the a_w of the environment and subsequently would have to remain quiescent, as shown in Figure 4.4. This value of 0.95 is also where the volume of tissue liquid gained by passive equilibrium processes is sufficient to sustain vital life functions needed in organ system development. If environmental water activity values go above a_w = 0.97, the rate of water uptake is greater than predicted by passive equilibrium conditions with environmental a_w and represents another water uptake mechanism. This latter mechanism is an active transport system and requires the coupling of energy with ion transport. The movement of free water is coupled with transit of monovalent ions, sodium, and potassium.

The free water requirements for larval and adult stages of brine shrimp living in various environmental salinities

TABLE 4.1 Water Activity, a_w, of Saturated Salt Solutions and Equilibrium Hydration Levels for Individuals of $CaSO_4$-dried *A. salina* Cysts Incubated in the Vapors of NaCl Solutions

	Water Activity a_w (P/P$_0$)	Individuals Measured (μl H_2O/cyst $\times$ 10^3 ± S.D.) ($n = 20$)
Saturated Salt Solutions at 29°C		
LiCl	0.116	2.247 ± 0.223
$Na_2Cr_2O_7$	0.525	1.619 ± 0.301
NaCl	0.755	1.284 ± 0.143
KCl	0.845	0.992 ± 0.218
KNO_3	0.913	0.901 ± 0.138
NaCl Solutions at 23°C		
88 g/l	0.95	.
117 g/l	0.93	
176 g/l	0.89	
234 g/l	0.35	
293 g/l	0.31	

SOURCES: Winston and Bates, 1960; Rockland, 1960; Lang, 1967; Clegg, 1976a.

have been established by the studies of Croghan (1958) and Conte et al. (1972, 1977). Brine shrimp maintain an uptake of water by metabolically controlling osmotic desiccation. Osmotic desiccation is caused by the loss of free water through semipermeable membranes surrounding living

TABLE 4.2 Metabolic Activities in Cysts as a Function of Hydration Level

NaCl Solution (g/l)	Hydration[a] (g H_2O/g dry weight cysts)	Metabolic Status[b]
73	0.65	Conventional metabolism, same events occur as in fully hydrated cysts in the presence of oxygen. In the absence of oxygen, no carbohydrate metabolism occurs, slow catabolism of diguanosine-tetraphosphate[c], pH_i decreases by >1 unit[d]
117	0.50	No respiration, some amino acid metabolism, decrease in glycogen
205	0.42-0.33	No data
293	0.28	Decrease in active transport system

[a]Clegg (1974, 1976a).
[b]Clegg (1978).
[c]Stocco et al. (1972).
[d]Busa et al. (1982).

organisms. This outward loss of water is offset by coupling the passive inward diffusion of ions carrying water with selective removal of sodium and chloride ions (Conte, 1977). Figure 4.5 compares the osmotic potential of changing environmental salinities in terms of the movement of free water molecules being translocated across a membrane along with each ion transported into the

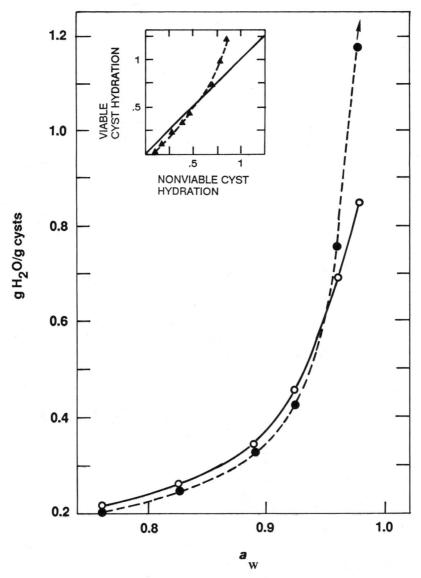

FIGURE 4.4 A comparison of the hydration behavior of viable (•) and nonviable (o) cysts. The cysts were rendered nonviable by exposure to ammonia vapors. The inset shows the relationship between the hydration of viable and nonviable cysts at specific water activities (a_w) (Clegg, 1978).

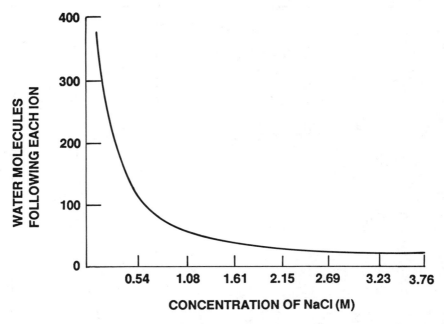

FIGURE 4.5 Water transport per ion translocated across the membrane as a function of solute concentration (Conte, 1977).

organism. The bioenergetic cost for maintaining the coupled ion-water exchange increases dramatically when an environmental water activity of 0.93 (equivalent to a salinity of 117 g NaCl/l) is reached (Conte, 1984).

In summary, the lower limit of environmental water activities (a_w) appears to be 0.95 for creating the internal fluid level of 0.65 g H_2O/g dry weight needed for resumption of vital cellular processes of brine shrimp. The effects of embryonic hydration on the other biological parameters are shown in Table 4.2.

Zooplankton

Reproduction and Development. The dynamics of larval brine shrimp populations can be strongly influenced by the physiological mechanisms that control the development of individual embryos. The most influential physical factors are salinity and temperature.

The Mono Lake brine shrimp species, *Artemia monica*, is diploid and bisexual. Fertilization occurs when sperm enter the ovisac after copulation. Work on strains of brine shrimp from San Francisco and Utah indicates that sperm are not stored within females (Bowen, 1962). Subsequently, cleavage of the zygote begins, and embryonic development can proceed along either of two lines. The female brine shrimp can retain the developing embryos within the ovisac and give birth to live, fully swimming naupliar larvae (the ovoviviparous pathway), or the female can interrupt embryogenesis by covering the embryos with a thick protein shell that inhibits development and then release dormant embryos as cysts (oviparous pathway). The control of the maternal shell gland and the factors that make the shell "impermeable," which allow the formation of a developmentally arrested embryo (cyst), are not completely understood. However, in some cysts of *A. monica* the breaking of dormancy appears to be triggered at release, since nauplii are produced shortly after cysts enter the water. In other circumstances, the cysts need to be "activated" before dormancy is broken. The activation process appears to depend upon the amount of free water and oxygen in the environment. Because the cysts of *A. monica*, unlike those of other brine shrimp, sink (Dana and Lenz, 1986), factors influencing environmental conditions near the bottom are of critical importance in the activation of these cysts.

Embryos that have become dormant because of developmental inhibition live in one of two distinct conditions: a diapause state or a quiescent state. Embryos under diapause conditions are kept dormant by endogenous factors (Drinkwater and Crowe, 1986) that prevent further resumption of development even under favorable environmental conditions. The diapaused embryo requires an activation stimulus that releases the embryo from the endogenous constraints and allows favorable environmental conditions to initiate the resumption of development. The quiescent embryo has usually been delayed by simple environmental constraints (lack of water, lack of oxygen, or low temperature) that prevent development. It does not require any auxiliary activation signal.   In ephemeral saline lakes, osmotic desiccation and anoxia are examples of environmental conditions found that produce quiescence.   When environmental conditions become satisfactory, such that the quiescent embryo can be fully hydrated and oxygenated, quiescence is usually broken and the embryo resumes full development to hatching.

Mono Lake, being a large, deep salt lake, does not fluctuate widely in salinity or dry up completely.   Therefore, environmental conditions under which cyst development and termination occur are unlike those found in shallow saline lakes.   For instance, because cysts from *A. monica* sink, they are always under conditions of hydration.   If cysts reside on the lake bottom in deep water, they have the additional environmental condition of remaining in a state of anoxia.   Thus, such cysts will never hatch unless the lake mixes down to the bottom.

Diapause Cysts and the Role of Intracellular pH (pHi).   The environmental requirements limiting the processes that control emergence (splitting of the shell) and hatching (escape of the nauplius) of *A. salina* are reviewed by Clegg and Conte (1980).   Recent studies of *A. monica* diapause cysts show no difference from *A. salina* in the role of pHi in activation of arrested embryos.   Previously published data (Busa et al., 1982; Busa and Crowe, 1983) had suggested that pHi in dormant cysts of *A. salina* was depressed. However, the cysts they used were not really in diapause; truly diapaused *A. salina* cysts, like those of *A. monica,* may have elevated pHi (Drinkwater and Crowe, 1987). Intracellular pH affected critical preemergent metabolic pathways, and pHi may be involved in the breaking of diapause in *A. salina* and in *A. monica* (Drinkwater and Crowe, 1987).

Diapause Cysts and the Role of Temperature.   An important feature of diapause regulation in *A. monica* is that cold creates conditions that allow typical preemergence carbohydrate and nucleotide metabolism to occur, which in other *Artemia* species is normally inhibited at lower temperatures.

Dana (1981) showed that release from diapause could be achieved by a long cold-hydration period (>90 days) at 5°C under anoxic conditions, and this release was not dependent on either a cycle of hydration/dehydration or a cycle of oxygenation (although oxygen is still required to end quiescence). The mechanism of this release from diapause is not understood. Thun and Starrett (1986) confirmed these findings and found that at 4°C and 90 g/l maximal hatching was achieved after 90 days. If cold-treated cysts were removed before the 90-day period and allowed to continue development at elevated temperatures (10°C) in the presence of oxygen at identical salinity, there was no increase in the number of nauplii that emerged or hatched. If the salinities of the media after the cold-hydration period were increased from 90 to 140 g/l and 90-day cold-treated cysts were reintroduced into these higher salinities, osmotic desiccation occurred and emergence and hatching of embryos diminished.

Drinkwater and Crowe (1986) reported that carbohydrate metabolism occurring during the cold-hydration period prepares *A. monica* embryos for preemergence in a manner similar to that reported for *A. franciscana* at elevated temperatures (Clegg, 1964). These biochemical changes include the breakdown of trehalose into glycerol, which normally occurs only under aerobic conditions and at higher temperatures in those cysts that are terminating diapause. In addition, cold-hydrated *A. monica* cysts can synthesize organic acids and glycogen at levels similar to those found in other species of *Artemia* at higher temperatures and oxygen levels (Conte et al., 1980). Therefore, it would seem that the Mono Lake brine shrimp has evolved controls that act differently from those of other *Artemia* species in avoiding the inhibiting effects of low temperatures on pre-emergence mechanisms.

Fluidity and Circulation. One functional prerequisite for the survival of brine shrimp is the formation of a circulatory system. The circulatory system provides a transport fluid containing essential nutrients and growth factors vital to the rapidly growing and differentiating embryonic cells. Additionally, the toxic metabolic wastes created by these cells are removed by the circulatory fluids because the fluids are constantly flushing the spaces surrounding the cells with large quantities of free water.

Drinkwater and Crowe (1986) have measured the maximal amount of bound water in *A. monica* cysts. In a comparison with other populations within the superspecies *A. franciscana*, both species showed that the population of cells in the cysts contained 0.297 g H_2O/g dry weight of cells. This is the limiting value of the water of hydration that protects the viability of brine shrimp cells. One can predict that increasing the amount of fluidity, by increasing the content of free water within tissue and cellular compartments up to a value of 0.60 g H_2O/g dry weight of cells, will allow restricted metabolism to occur and will allow the physiological functions of termination of diapause, gastrulation differentiation, and formation and preemergence of the nauplius. The mechanism for providing the entrance of free water in the restricted metabolism stage of the quiescent embryo is dependent upon internal glycerol gradients that offset osmotic desiccation. This mechanism remains successful as long as the membranes surrounding the embryo do not leak glycerol. Hatching of cysts does not occur unless the content of free water reaches 0.60 g H_2O/g dry weight of the cells.

Hatching destroys the glycerol-impermeable membranes and with it the passive mechanism of water balance in the growing embryo. A new active mechanism of ionic (Na^+) excretion via an enzymatic Na,K-ATPase membrane pump is initiated in the free swimming nauplius (Conte, 1984). It is this ATP-utilizing enzymatic mechanism that transports water into the nauplius against the osmotic gradient. This sodium pump provides the additional free water that accounts for the elevation of embryonic water content (>1.2 g H_2O/g dry weight) above a_w values predicted from equilibrium measurements of the saline medium (Clegg, 1978). The increase in prenaupliar water content provides for more tissue fluidity and allows unrestricted oxidative metabolism. In turn, naupliar circulation and transport of food substrates are enhanced, furnishing optimal conditions for the initiation of other physiological activities, such as locomotion, feeding, and digestion.

Enzymatic assays for the existence of the sodium pump in *A. monica* have not been made. However, indirect evidence for the existence of a sodium pump in *A. monica* nauplii and adults comes from the studies by Dana and Lenz (1986) during larval growth at various salinities of Mono Lake water. Their findings, based upon survival, growth, and the development of the nauplius to the juvenile stage, show that salinities above 133 g/l (TDS) significantly depress these life processes. Nauplii living in lower salinities were affected less and were capable of continued postabdominal development of swimming legs and subsequent metamorphosis into juveniles and adults. Since feeding of nauplii occurred as part of the experimental treatment, it would appear that food capture, assimilation, and digestion were not adversely affected.

Bioenergetics and Fecundity. The metabolism of food provides the developing embryo with the energy necessary to balance the various energy-utilizing reactions. The production of a circulatory fluid having a low salt content requires distillation of ingested brines, a process that has a high energetic cost.

Dana and Lenz (1986) evaluated the reproductive potential of laboratory-reared *A. monica*, Mono Basin brine shrimp, with unlimited food supply. Young reproductive adults were paired to assess the effect of salinity on reproduction. The number of eggs per brood and hatching rate of cysts appeared to be reduced at salinities above 118 g/l (TDS). Low hatching was observed at a salinity of 133 g/l, and no hatching at a salinity of 159 g/l. Reproductive potential as judged by egg formation, egg fertilization, egg number, mode of development of fertilized egg, and female mortality appears to be adequate until increased salinity (osmotic desiccation) lowers water content (a_w) below 0.95 g H_2O/g dry weight. Above these salinities, the amount of energy available for reproduction is reduced because egg formation and development are lower priorities than the need for free water for maintaining the circulatory system.

Zoobenthic Organisms

Reproduction and Development.   The principal species of
zoobenthos at Mono Lake is the brine fly, *Ephydra hians*.
The population is diploid and bisexual, as are other populations of ephydrids found in the Great Salt Lake and Great
Basin drainage area (Collins, 1975, 1980a,b).   The female
has a spermatheca for storing sperm after copulation.   The
female provides special nutrients for the oocytes via acces-
sory nurse cells, allowing vitellogenesis, the deposition of
yolk in the lower parts of the ovariole, and a very large
egg.   She also provides a chorion for attaching the eggs to
the hard substrates where they are laid.   The vitelline
membrane surrounds the outer layer of the oocyte and is
laid down at the end of vitellogenesis.   The membrane
serves as the protective barrier between the environment
and embryo.

Oviposition (egg-laying) behavior by female brine flies
influences their spatial distribution.   The newly laid, fertilized egg is a hydrated embryo covered with a nonsticky,
opalescent, semipermeable shell.   If the egg is not kept
fully hydrated, it will die.   Therefore, the selection by the
female of a suitable benthic site for the egg is of critical
importance.   The embryo must be adequately protected by
its environment, and a food source must be available to the
larva when it emerges.   The ovipositing female must also
be able to breathe while she is underwater.   The female fly
obtains the needed oxygen by forming an air bubble that
surrounds most of her body.   The deposited eggs adhere to
the substratum surface.   When in place, the deposited egg
is continuously bathed by the saline water, which furnishes
oxygen while removing waste from the growing embryo.
Most importantly, embryonic growth processes can proceed
only if the hydration state of the early fertilized egg is
maintained.   The osmoregulatory mechanisms used by the
fertilized egg to combat osmotic desiccation are unknown.

Fluidity and Circulation.   Loss of free water by osmotic
desiccation in early brine fly embryogenesis results in increased mortality of adult and juvenile life stages.   Herbst
(1981) found that Mono Lake water at a salinity of 120 g/l

(TDS) would cause larval mortality and prolong time of development of first, second, and third instars. The pathology of saline-induced larval death is not known. It may be assumed that the lack of circulating fluids for growth processes impairs development.

Bioenergetics. The embryo of the brine fly uses stored yolk material for growth up to the first instar, after which the larva begins to graze upon the microbial-algal mat covering the substrate. Herbst (1986) found that a reduction of 50 percent of "normal" food rations significantly inhibited larval development beyond the third instar and prevented pupation and adult emergence. Development was best when larvae were fed microbial-algal-diatom mixtures rather than a single-algal diet. Most life history traits of the brine fly are adversely affected as osmotic stress of the environment increases, especially when the salinities exceed 120 to 130 g/l TDS. The detrimental effect of osmotic stress is related to the energy debt created by excretory organs working against the osmotic constraints. The energy required by the excretory organs to maintain osmotic regulation reduces the total energy available for the growth and development of any life stage.

Adaptation of Brine Shrimp and Brine Fly to Changes in Salinity

If brine flies and brine shrimp could adapt genetically to increased salinity and alkalinity, a lowered lake level would have a smaller effect on their populations. There are two lines of evidence suggesting that the probability that brine flies and brine shrimp would adapt to higher salinities (>150 g/l) is remote. First, although alkaline, saline lakes in North America, Africa, Asia, and Australia with salinities greater than 150 g/l support macroinvertebrate populations, none are as alkaline as Mono Lake. The combination of high alkalinity and salinity appears to be very difficult to adapt to. Second, whenever it has been examined, the stoichiometry of the sodium pump, which is the critical mechanism that organisms use to maintain adequate free water against strong osmotic gradients, is the same for both vertebrates and invertebrates. Thus it appears that in its function, at least, the sodium pump has little or no variability available on which natural selection can act.

BIRD POPULATIONS: SECONDARY CONSUMERS

At saline and alkaline lakes, birds are often conspicuous top consumers in relatively short food chains. Birds are attracted to lakes such as Mono because they provide an unusual abundance of food. The harsh chemical conditions of the lake preclude the existence there of many kinds of aquatic predators, particularly fish. Without aquatic predators, populations of some aquatic invertebrates reach extraordinarily high densities. Nonaquatic predators, such as birds that exploit these populations, show typical numerical and functional responses to the higher concentrations of prey. Impressive concentrations of birds can result, as at Mono Lake, where several species exploit the lake's production of brine shrimp and brine fly.

Some bird species exploit these abundant invertebrate populations facultatively. In other cases birds may have a more obligatory reliance on the abundant invertebrate populations as a crucial food source during portions of their annual cycle. If these food resources were diminished or lost and alternative sources of food could not compensate, dependent bird populations would probably experience major limitations at the crucial stage in their life cycle when they formerly relied on the lake's resources.

At Mono Lake, three species--eared grebes, Wilson's phalaropes, and California gulls--may have a more or less obligatory dependence on the lake's seasonally abundant invertebrates. Each of these three species is present in extraordinarily high numbers, and each exploits brine shrimp and brine fly populations. The grebes and phalaropes rely on Mono Lake's aquatic resources during stopovers on their fall migrations. The gulls nest on islands in Mono Lake. Abundant food is, however, a common attraction for these birds as well as others that visit the lake in smaller numbers.   The numbers of eared grebes, Wilson's phalaropes, and California gulls that visit Mono Lake represent substantial proportions of the North American populations of each species.   Our main challenge is to predict the responses of these bird populations to a possible reduction in their food supply at Mono Lake.   If these birds are dependent on Mono Lake's food resources, changes in the lake's aquatic community could have major consequences for North American populations of all three species.   If, on the other hand, the birds could switch to alternative food supplies at other locations, they could, in the long run, show only minor disruptions in the normal dynamics of their populations.

Because of the central importance of understanding the relationships between Mono Lake and the eared grebes, Wilson's phalaropes, and California gulls that visit it, separate, detailed accounts of each of these species are presented.   In these accounts, particular attention is paid to the estimated numbers of individuals that visit Mono Lake and their patterns of habitat and resource utilization while at the lake.   On the basis of the population data available, an attempt is made to determine the relationships between the birds visiting Mono Lake and the rest of their regional and continental populations.

Brief discussions of the other aquatic bird species that are present in smaller numbers or that use Mono Lake's resources more opportunistically and the nonaquatic birds that visit the Mono Basin are presented in chapter 5.

Eared Grebes

Eared grebes (*Podiceps nigricollis*) use Mono Lake primarily as a stopover site during fall migration; much smaller numbers of grebes use the lake in other seasons.   In North America, only Great Salt Lake and the Salton Sea attract as many grebes as Mono Lake.   The seasonal cycle of grebe occurrence at the lake is shown in Figure 4.6. Peak numbers of grebes occur during late September, October, and November (Winkler, 1977; Jehl, 1982a; Cooper et al., 1984), a time when grebe predation contributes to the annual decline in brine shrimp densities (Cooper et al.,

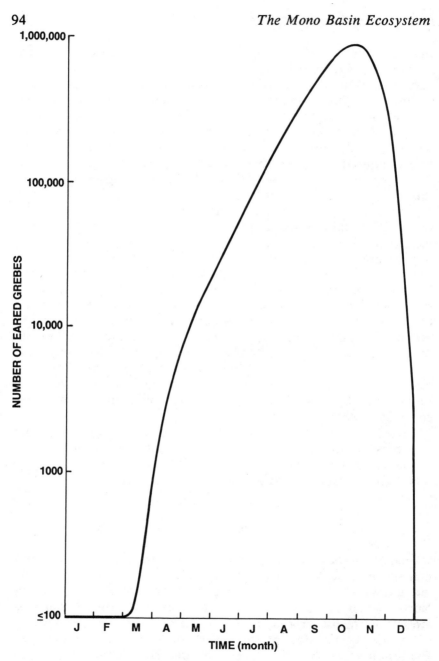

FIGURE 4.6 Estimates of the number of eared grebes on Mono Lake each month in a typical year (based on Winkler, 1977; Jehl, 1982a; Cooper et al., 1984).

1984). The departure of grebes from Mono Lake is precipitated by the seasonal collapse of the lake's population of adult brine shrimp. Grebes would probably prolong their stay at Mono Lake, perhaps even overwintering as they do in the Salton Sea, if their food remained abundant.

Annual estimates of peak grebe numbers are usually in the range of 600,000 to 900,000 (Jehl, 1982a; Cooper et al., 1984). There are, however, major difficulties in arriving at a precise estimate of these peak numbers. Although census methods have improved since 1976, when the earliest systematic efforts took place, they still fail to provide convincing estimates of variances. Confidence intervals around the population estimates remain unknown, making statistical comparisons between time periods difficult. Furthermore, estimates of peak numbers do not necessarily reflect the total number of grebes that visit the lake during the year. There is certain to be some turnover of grebes during the fall, so that total numbers exceed peak numbers by some unknown percentage. Information on individually recognizable birds suggests, however, that turnover rates may be relatively low under present lake conditions (Jehl, 1983c). The origins and destinations of the migrant grebes that visit Mono Lake each fall are not well known, and at present there are too few banding data to clarify the issue (Jehl and Yochem, 1986).

Estimates of the size of the entire grebe population in North America are much less precise than estimates of grebe numbers at Mono Lake. Totaling the more or less contemporaneous counts of grebes at major fall migration stopovers, J. R. Jehl, Jr. (Hubbs-Sea World Research Institute, personal communication) estimates a continent-wide fall population of over 2.5 million, but there is no way of knowing what proportion of these birds are counted more than once as they move between the major concentration points. Fewer estimates are available for the scattered birds that do not visit the major stopover sites. Although these population figures certainly leave much to be desired, the conclusion that Mono Lake is visited by one quarter to one third of the eared grebes in North America each year seems to be justified.

While at Mono Lake, grebes feed on adult brine shrimp and brine fly larvae, these two food items constituting over

95 percent of their diet from August to November (Winkler and Cooper, 1986). Ingestion of salt while feeding on these invertebrates is not a physiological problem for grebes, nor is it likely to be over the range of salinities that brine shrimp can tolerate (Mahoney and Jehl, 1985). Grebes swallow minimal amounts of saltwater while ingesting their food and apparently can meet their water requirements with water obtained from their food.

Eared grebes are gregarious migratory birds; they typically concentrate at a few major stopover sites, especially during the autumn migration (Palmer, 1962; Cramp and Simmons, 1977). While at these autumn stopover sites, grebes gain weight and, in some instances, undergo a more or less extensive molt. It seems that the degree to which an extensive molt takes place is correlated with the abundance of food at a particular stopover site. When food is very abundant, as it is at Mono Lake, the grebes remain at the stopover site for an extended period of time and undergo a major molt in which they replace most of their body feathers in about 6 weeks (Storer and Jehl, 1984). In contrast, at less productive sites, the stopover may be shorter and the molt less extensive. It is likely that massive, rapid molting is triggered by the birds reaching a threshold in general body condition; J. R. Jehl, Jr. (personal communication) has suggested this threshold may be about 50 g above the weight on arrival at a stopover point. If the bird fails to reach this threshold while at a stopover site, the molt will be delayed until the wintering area is reached.

The body weight of a grebe is a good indicator of the amount of fat the bird has accumulated (Winkler and Cooper, 1986), and the weight of its fat reserves determines its flight range. While visiting Mono Lake, all eared grebes gain weight--some may more than double their weight--by adding body fat. Jehl (1982a) reports that body weight of starving grebes averages about 250 g; this suggests an average fat-free weight of about 225 g because 10 percent of a bird's body weight represents unusable fat reserves (Blem, 1980). The average grebe probably must weigh at least 275 g in order to undertake a long-distance flight. The weights of some of the earliest migrants to arrive at Mono Lake, particularly of juveniles, indicate they have essentially no usable fat reserves. They may be in such poor condition that they would be incapable of successfully extending their migration very far beyond Mono Lake. Birds arriving later weigh more (>385 g), having accumulated fat earlier in their migration. Winkler and Cooper (1986) report that peak weight for eared grebes at Mono Lake reaches about 500 g. Birds of this weight have usable fat reserves far in excess of what is required for a typical migratory flight to wintering areas.

The Salton Sea, the Pacific coast, and the Gulf of California--areas where many grebes spend the winter (Palmer, 1962)--are within reach of grebes leaving Mono Lake with at least 45 to 85 g of usable fat (Figure 4.7). The majority of the grebes that visit Mono Lake do not, therefore, need to become as fat as they do to successfully reach wintering areas, but deprived of food resources of Mono Lake, an unknown but smaller number of grebes would have difficulty continuing their migration. It is not known whether or not food resources during the fall in the Salton Sea, Pacific Ocean, and Gulf of California would be able to support all of the grebes that visit Mono Lake if they visited these areas earlier in the autumn than they do now.

The primary feature of Mono Lake of importance to eared grebes is the abundance of brine shrimp. During August through October, eared grebes gain weight at a rate of about 3 g/day (Jehl, 1982a), and Cooper et al. (1984) estimate a consumption rate of up to 70,000 shrimp/day; these crude figures are, however, based on population averages and not measurements of individual weight gains or consumption rates. The critical density of brine shrimp below which grebes would completely abandon Mono Lake as a stopover site is difficult to predict, but the yearly exodus of birds coincides approximately with the time when mean lakewide densities of brine shrimp drop below about 25,000/m^2 (Cooper et al., 1984). Furthermore, the distribution of grebes over the lake's surface is patchy, perhaps reflecting local variations in brine shrimp densities (Lenz et al., 1986). Grebes are particularly concentrated in regions of the lake with brine shrimp densities over 20,000/m^2. Both of these independent measures of the association between grebes and brine shrimp densities suggest a threshold density of about 20,000 to 25,000 shrimp/m^2 below which grebes may find brine shrimp to be a

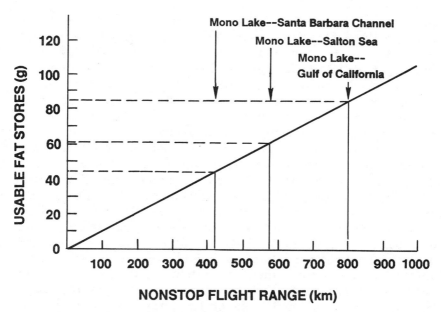

FIGURE 4.7 The potential nonstop flight range of an eared grebe as a function of usable fat reserves. Calculated using the most conservative estimates available for migration energetics (9.0 kcal/g of fat, 1.05 kcal/k of nonstop flight) according to Blem (1980).

marginally profitable food on which they could not easily gain weight. These measurements of shrimp abundance refer to the number of shrimp caught with a plankton net lifted vertically over the entire water column. Most samples collected from discrete depths in early autumn contain 1 to 10 brine shrimp per liter in the upper 10 to 15 m (Lenz, 1980; Conte et al., in press).

One might, therefore, expect that grebes arriving in good body condition (with >45 to 85 g of usable fat) would not stay long at Mono Lake or molt heavily if mean lakewide brine shrimp densities were below 25,000/m^2 or if similarly high local concentrations of shrimp could not be located. On the other hand, grebes arriving in poorer condition might stay for a while and attempt to slowly fatten themselves, even if brine shrimp densities were lower. Grebes are known to stop at lakes with invertebrate densities below those at which they leave Mono Lake (Winkler, 1982), but at present we cannot predict with certainty the critical density of brine shrimp below which it becomes difficult for a grebe to gain weight.

If the invertebrate populations of Mono Lake were to decline, the committee hypothesizes the following sequence of responses by the eared grebes. As invertebrate numbers began to drop below recent levels, grebes would continue to visit Mono Lake in current numbers, but they would not stay as long as they do now because they would deplete shrimp populations earlier in the season. The lower the number of shrimp became, the sooner grebes would depart. At some point, the shortened stay at Mono Lake would be incompatible with the type of rapid, heavy molting that now occurs. Grebes would then be forced to interrupt or delay their molt until they reached wintering areas, as they apparently do in eastern North America (Palmer, 1962). If brine shrimp densities reached a level so low that it would be impossible for grebes to rapidly fatten themselves (probably somewhere below 20,000 shrimp/m^2), individuals arriving at Mono Lake with sufficient fat reserves to continue migrating would probably do so. Only grebes arriving in relatively poor condition would remain, but only if Mono Lake's food resources were superior to those offered by other lakes in the Great Basin. At this level of food availability, the total number of grebes visiting the lake during the fall might remain essentially unchanged from current numbers, but turnover rates would be high and numbers present at any one time would be much lower than under present conditions. When Mono Lake's invertebrate populations became no different from those at other potential stopover sites, grebes would eventually redistribute themselves among other lakes, perhaps occupying lakes that currently receive little or no use (Winkler, 1982) because their food resources compare unfavorably with Mono Lake's.

The impact that this hypothesized sequence of events would have on the dynamics of the North American eared grebe population can only be roughly assessed, but speculations in the popular press about a possible devastation of the grebe population seem unwarranted. Shortening the length of stay at Mono Lake and interrupting or delaying the molt would probably not result in a rise in mortality rates. When grebes began to leave Mono Lake with fat reserves that were only marginally above the minimal levels needed to reach wintering areas, mortality rates could be expected to rise. Migrating grebes can be grounded by inclement weather, and extra fat resources would be needed to sustain the birds during these crises. Lacking these extra reserves, a portion of the population would, each year, be vulnerable to unpredictable but catastrophic losses, such as those described by Jehl and Bond (1983). The impact on the North American population would probably be manifested in greater year-to-year variations in population size than occur now. If grebes were largely redistributed among other stopover sites when Mono Lake's food resources became comparable to those at other lakes, there might be a reduction in the North American population, but its magnitude is impossible to predict. The eventual population might be determined by the unknown carrying capacities of lakes that are, for the most part, not now supporting anywhere near the numbers of grebes that they might be capable of supporting because birds are so concentrated at Mono Lake. Under these circumstances, the autumn migration period could potentially become the limiting season of the year for grebes. Mono Lake's rich food resources have so far kept this from happening. Mortality rates among grebes at Mono Lake are low (Jehl, 1981b), and major population limitations are probably now associated with the grebes' wintering or breeding areas.

Phalaropes

Three species of phalaropes use Mono Lake during migration, Wilson's phalarope (*Phalaropus tricolor*), red phalarope (*Phalaropus fulicarius*), and red-necked or northern phalarope (*Phalaropus lobatus*). Red phalaropes visit Mono Lake in very small numbers (Jehl recorded three in 3 years) during their fall migration (Jehl, 1986), and essentially nothing is recorded of their use of the lake.

Moderate numbers of red-necked phalaropes use Mono Lake as a brief migratory stopover. Jehl (1986) estimates the total population using the lake each year to be 2.5

times his maximum count for that year, although the committee feels that this multiplier is inadequately justified. Additionally, his estimates have no confidence intervals, which are needed to assess the statistical significance of year-to-year changes in numbers. However, accepting his figures, 43,000 red-necked phalaropes used Mono Lake in 1981, 54,000 in 1982, and 33,000 in 1983. There is no reason to suppose that these apparent changes in numbers reflect events at Mono Lake. Jehl speculates that the 1983 drop in numbers reflects mortality on the oceanic wintering grounds due to the 1982 El Niño. Other year-to-year variations may reflect the use of alternative lakes as stopping points, but data are inadequate to test this hypothesis. There are no data useful for estimating the portion of the North American population of red-necked phalaropes using Mono Lake, but it is probably fairly small given the numbers using other western saline lakes (Jehl, 1986). However, of the Great Basin lakes in northern California, Mono Lake supports one of the highest peak populations of red-necked phalaropes, and is thus of local importance (Winkler, 1982; J. R. Jehl, Jr., personal communication, 1986).

Wilson's phalaropes are the most abundant phalarope using Mono Lake, with peak populations of 93,000 to 100,000 or more (Winkler, 1982; Mahoney and Jehl, 1984). J. R. Jehl, Jr. (personal communication, 1986) estimates a peak of 70,000 with a total population use of 100,000 to 125,000. Unlike red-necked phalaropes, Wilson's phalaropes make an extended stop at Mono Lake, using it as a staging area before commencing what may be a nonstop migration to South America (Jehl, 1981a). While at Mono Lake they complete their molt and put on fat. It is unknown what percentage of the North American population uses Mono Lake, but Jehl (1981a) suggests that Mono Lake is the largest staging area for Wilson's phalaropes migrating through western North America. Certainly, Mono Lake is of major importance to this species in California (Winkler, 1982).

The time of arrival of phalaropes at Mono Lake coincides with the presence of immense numbers of brine shrimp and brine flies. Post-breeding red-necked phalaropes begin to arrive at Mono Lake in early to mid-July (Jehl, 1981c, 1982b, 1986). Numbers build until early

August, remain high until early September, and then drop steadily until most birds are gone in early or mid-October. Wilson's phalaropes begin to arrive in middle to late June. After a period of 35 to 40 days needed to replace body plumage and lay on fat, the birds begin their southward migration near the end of July, with females preceding males (Jehl, 1981a, personal communication). Wilson's phalaropes are gone from the lake by September.

The principal prey of red-necked phalaropes at Mono Lake is the brine fly (Jehl, 1986). The phalaropes take adult flies, as well as pupae and larvae; they will eat brine shrimp when brine flies are not available. Red-necked phalaropes at Great Salt Lake also apparently prefer brine flies to brine shrimp (Wetmore, 1925). Brine flies are likely preferred to brine shrimp because of their larger size and presumed higher nutritional content. There are no data concerning critical densities of prey needed, nor is there an estimate of prey consumption by red-necked phalaropes at Mono Lake.

The Wilson's phalarope at Mono Lake takes both brine shrimp and brine flies (Jehl, 1981a; Mahoney and Jehl, 1984). There is, however, a surprising lack of quantitative information on the diet of Wilson's phalaropes at Mono Lake, particularly in light of the importance of the lake as a migratory staging area for this species. Winkler (1977) reports on the examination of eight specimens from Mono Lake, which contained 93 percent brine flies and 7 percent brine shrimp. There are insufficient data available to estimate the total food consumption of Wilson's phalaropes and the partitioning of diet between prey species at Mono Lake.

At Mono Lake, red-necked phalaropes concentrate over shallow submerged rock formations and submerged vegetation mats near shore where pupating brine flies are abundant, such as near tufa towers and the smaller islands (Jehl, 1981c, 1982b, 1986). Late in the season, possibly in preparation for migration, birds congregate in small flocks at the center of the lake.

Access to fresh water may not be required by red-necked phalaropes, but they visit freshwater sources around the lake to drink and bathe (Jehl, 1986). Although data on the physiological tolerances of red-necked phalaropes to salt and alkali concentrations are lacking, it is reasonable to assume that Mono Lake water exceeds their tolerances for salt ingestion (Mahoney and Jehl, 1984, 1985). As is true for the other Mono Lake birds, the red-necked phalaropes undoubtedly avoid salt loading by taking hyposmotic prey and by minimizing their intake of lake water.

Habitat use by Wilson's phalaropes at Mono Lake has not been described in detail. However, Wilson's phalaropes make extensive use of emergent tufa towers ˏand nearshore sandbars for roosting (J. R. Jehl, Jr., personal communication) and they forage over much of the lake's surface for brine shrimp and brine flies. Given the virtual absence of any quantitative data on this species at Mono Lake, it is difficult to evaluate habitat requirements, except that it is clear that the lake provides a foraging habitat and a location for molting while staging in preparation for migration to South America (Jehl, 1986).

Wilson's phalaropes at Mono Lake apparently avoid incurring a salt load by minimizing their ingestion of Mono Lake water and by taking hyposmotic prey (Mahoney and Jehl, 1984). Just before migration, when feeding rates are maximized for fat accumulation, Wilson's phalaropes at Mono Lake show enlarged salt glands and at this time visit freshwater sources morning and evening (Mahoney and Jehl, 1984). These data suggest that the availability of fresh water for 2 weeks prior to migration from the lake may be important for coping with excess salts ingested during the period of heavy foraging associated with fat accumulation.

The use of Mono Lake as a staging area by Wilson's phalaropes has important implications for the population involved. The use of staging areas is often traditional in shorebirds (Pitelka, 1979), and most individuals show great fidelity to a given site once a migration pattern is established (Pienkowski and Evans, 1985; Evans and Townshend, in press). Adults with site attachment might be reluctant to leave a staging area such as Mono Lake before molting and putting on fat reserves. Many shorebirds arrive at staging areas with relatively little reserve fat and in an exhausted condition (Dick and Pienkowski, 1978). Weight data from Wilson's phalaropes arriving at Mono Lake show low, but not depleted fat reserves (J. R. Jehl, Jr., personal communication). Additionally, Jehl (personal communication) has some data that indicate that Wilson's phalaropes at Mono Lake may be able to shift to alternative sites. At present it is difficult to predict the response of adult Wilson's phalaropes if they arrived at Mono Lake and found their expected food sources severely diminished, or the population consequences if they attempted to use alternate sites for staging.

California Gulls

California gulls (*Larus californicus*) nest on several of the islands and islets in Mono Lake. In addition, California gulls nesting elsewhere in the Great Basin may visit Mono Lake during migration (J. R. Jehl, Jr., personal communication). A variety of reports and publications provide estimates of the number of California gulls nesting at Mono Lake (Winkler, 1979, 1983a; Power et al., 1980; Jehl, 1983a,b, 1984, 1985; Jehl et al., 1984; Shuford et al., 1984, 1985; Shuford, 1985; D. Winkler and W. D. Shuford, Cornell University, and Point Reyes Bird Observatory, respectively, personal communication). The extent of use of Mono Lake by California gulls in migration is not known.

The methods for assessing the numbers of California gulls nesting at Mono Lake have varied considerably over historical times. Early records of numbers are generally based on crude estimates by naturalists or even on hearsay (Jehl et al., 1984; D. Winkler and W. D. Shuford, personal communication). Since 1979, observers have estimated breeding populations of gulls based on counts of chicks seen in the colonies multiplied by correction factors for chicks produced per pair and the visibility of chicks on each colony. The accuracy of this census method has been questioned by Jehl (1983a, 1984, 1985), as it has produced variations from his counts of up to 170 percent for some islets (Jehl, 1985).

The population of California gulls nesting at Mono Lake has increased from an historical low of 3,000 to 4,000 birds in the early 1900s to 50,000 by 1976, since when it has been relatively constant at between 40,000 and 50,000 birds (Jehl et al., 1984; Shuford, 1985). Jehl et al. (1984) thought that the population was more or less constant in size from 1916 to the early 1950s, after which they hypothesized a rapid expansion of the population, driven by immigration. According to their view, the provision of new nest sites by the lowering lake level made the population increase possible. In contrast, D. Winkler and W. D. Shuford (personal communication) believe that the Mono Lake gull population has been recovering from excess egg collecting in the 1800s and has undergone an exponential increase to its former numbers. In their estimation, the gull population in the past has not been constrained by insufficient nesting space. The California gull nests on islands in lakes throughout much of the western United States and central Canada (Power et al., 1980). Despite variation between California gulls on Mono Lake and some other populations (Winkler, 1977, 1985; J. R. Jehl, Jr., personal communication), there is no evidence for treating the Mono Basin population as a separate local race or subspecies (Zink and Winkler, 1983). Several authors have estimated the significance of the Mono Lake California gull population in relation to regional and North American (world) populations of the species. Power et al. (1980) estimated a world population of about 220,000 California gulls, of which the Mono Lake population of about 47,000 in 1976 to 1978 made up 21 percent. However, this species is expanding its population in California and in the Great Basin (Conover and Conover, 1981; Conover, 1983), and estimates of the relative importance of various colonies or subpopulations are quickly out of date. Given the lack of precision in estimates of the world population, the Mono Lake population is likely to be between 15 and 25 percent of the breeding population of California gulls. The Mono Lake colony is the second largest known California gull colony in North America (Conover, 1983). From almost any perspective, the Mono Lake colony supports an important segment of this species' population.

Mono Lake is used by California gulls primarily between early April and early August. California gulls arrive at Mono Lake as early as the first week of March (Jehl, 1983b), with peak numbers arriving probably in early to mid-April (Power et al., 1980). Egg-laying begins in late April, peaks in mid-May, and is mostly completed by the first week of June (Jehl, 1983b; Shuford et al., 1985).

Chicks begin hatching in late May, hatching peaks in early to mid-June, and is essentially complete by the end of June (Shuford et al., 1985). Peak fledging occurred between July 10 and 21 in 1983 and 1984. California gulls begin to leave Mono Lake about mid-July, with a major portion of the population gone by mid-August (Power et al., 1980; J. R. Jehl, Jr., personal communication). Egg-laying takes place while brine shrimp are relatively scarce, but chick hatching and growth usually coincide with a period when this principal source of food is at or near peak abundance (Winkler, 1983b, 1985).

Several aspects of the breeding biology of California gulls at Mono Lake are potentially sensitive to changes in lake levels. The production of young per pair depends upon clutch size, hatching success, the number of chicks surviving to fledging at about 42 days of age, and freedom from disturbance (Power et al., 1980; Winkler, 1983b).

California gulls nesting at Mono Lake produce smaller eggs and smaller clutches than do birds in other colonies of this species (Winkler, 1985). Winkler (1985) argues that this smaller clutch size (and reduced egg volume) is caused by a scarcity of brine shrimp early in the season at the time of egg formation. Energy for egg production is apparently not sufficient for the female gulls to produce a clutch of three full-sized eggs. Adult gulls at Mono Lake have less food in their stomachs, and they courtship-feed less frequently during the pre-egg stage than gulls at Great Salt Lake, where clutches are larger (Winkler, 1985). Both of these findings support the hypothesis that food is limited early in the breeding season.

Hatching and fledging success of gulls at Mono Lake vary between years. Survival to hatching has been as low as 40 percent of eggs laid, as in 1982 (Winkler, 1983a,b), and in 1981 apparently fewer than 5 percent of chicks survived to fledging. In a "normal" year, between 30 and 35 percent of chicks fledge (Winkler, 1983b), and in a "good" year, up to 70 percent of chicks may survive (J. R. Jehl, Jr., personal communication).

Variations in methodology and incomplete data sets make year-to-year comparisons and estimates of chick production difficult. An average of about 0.5 chicks fledged per pair in the period 1983 through 1985. A portion of the interannual variation in chick production may be the result of infestation of chicks by ticks (Shuford, 1985). Post-fledging mortality of independent young prior to departure also varies greatly between years; in 1981 approximately 70 percent of fledged chicks died before migration, while in 1982 only about 2 percent were lost (Jehl, 1981b; Jehl and Jehl, 1981, 1982).

Year-to-year variability in the production of chicks is common among most species of gulls and by itself does not indicate a local population is likely to decline. In their lifetime a pair must produce but two young that survive to breed if a population is to remain stable. Power et al. (1980) have estimated, based on band returns, that most California gulls live about 9 years, with a few surviving to 13 years. These estimates may well be conservative, as band loss and its effect on the underestimation of age of survival were not considered (Kadlec and Drury, 1969). The average of 0.5 chicks fledged per pair in the period 1983 through 1985 may be sufficient for maintaining a stable or even moderately increasing population size, but we have insufficient data to know. The data base is currently very incomplete, particularly in the area of adult mortality rates, and the years sampled for chick productivity and post-fledging survival are few. Hence, it is premature to draw conclusions about the stability and growth potential of this population based on the reproductive and mortality data available.

California gulls at Mono Lake have three requirements of their habitat. They need predator- and disturbance-free areas on which to breed; they require a plentiful, nearby source of food; and they need access to fresh water. The first two requirements must be met in the immediate vicinity of Mono Lake if the present breeding population is to be maintained there. It is possible that alternate fresh-water sources within several tens of kilometers could be substituted for the present use of springs along the edge of the lake.

The importance of nesting areas free of predators and disturbance is underscored by the history of use of various islands in Mono Lake and changes in reproductive output on those islands. Paoha Island has been an important nesting area in the past but more recently was deserted either due to a population increase of feral goats or possibly due to coyotes (*Canis latrans*) (Jehl et al., 1984; D. Winkler and W. D. Shuford, personal communication). Although coyotes apparently did not disturb gull nesting on Negit Island in 1978, the first breeding season with a land bridge to the island, on July 10, 1978, canid scat and tracks were found on the island (D. Winkler and W. D. Shuford, personal communication). The next season (1979), breeding was disrupted: on May 12 and 13 Winkler (D. Winkler and W. D. Shuford, personal communication) found "normal" numbers of gulls nesting, but "on 27 May he found only scattered broken eggshells. No human tracks were seen, and the sandy soil of the colony site was crisscrossed with canid tracks." Given contradictory statements in Jehl et al. (1984) and D. Winkler and W. D. Shuford (personal communication), it is not clear whether birds deserting Negit Island settled elsewhere to breed in 1979 or simply did not breed that year. However, both sources agree that the Paoha Islets were used for the first time in 1979, and Jehl et al. (1984) state "while it is plausible that the arrival of coyotes prompted the gulls' relocation, we point out that there were no significant barriers to prevent the coyotes' arrival several years earlier." The fact that the canids took one or more years to discover the colony on Negit Island after the land bridge was formed is less important than the fact that the colony was broken up and deserted once canids arrived. Similarly, large colonies on Twain and Jarvis islets were deserted in 1982 after land bridges connected them to the mainland in 1981 (D. Winkler, personal communication). On the basis of experimental evidence, we know that breeding activities will be disrupted in gull colonies invaded by large terrestrial mammalian predators (Kadlec, 1971).

The type and amount of cover, whether vegetation, rock, or other material, may affect the desirability of a given area as nesting habitat. Pugesek and Diem (1983) studied habitat selection by California gulls nesting in Wyoming and found that gulls nesting in shrubbery had higher nesting success than those nesting in the open. Gull chicks can die from overheating, and in some cases vegetation or rock crevices provide protection from insolation (Salzman, 1982; Chappell et al., 1984). On the small islets, temperatures frequently exceed the thermoregulatory abilities of chicks, and shading by parent gulls appears critical for chick survival (Chappell et al., 1984). Dense vegetation also may block cooling breezes, and on Negit Island temperatures near the ground in areas of dense vegetation were higher than those found on Coyote Islet in the open near the shore (Jehl and Mahoney, in press).

Cover in some cases may be important for providing protection from avian predators, and at Mono Lake, great horned owl (*Bubo virginianus*) predation apparently caused the gulls to abandon nesting on several small unvegetated islets in 1982 and 1983 (Jehl et al., 1984). However, in other instances, gulls nesting in dense cover may be more susceptible to predation if predators can approach undetected, or if gull escape is impaired (Jehl and Chase, in press).

Gulls are generalist foragers and will take a wide variety of foods. California gulls are no exception; they eat many types of natural prey and foods obtained from humans (Power et al., 1980). Early reports of prey of California gulls at Mono Lake include brine fly larvae (Nichols, 1938; Young, 1952), trout (Nichols, 1938), and brine shrimp (Young, 1952; Winkler, 1977). More recently, Jehl and Mahoney (1983) reported on prey found in 18 adults, 73 chicks, and 20 fledglings in 1982. The vast majority of the prey was brine shrimp, with small amounts of brine flies, fish, and freshwater insects. Garbage had been eaten as well. Substantial numbers of cicadas (no taxon given) were eaten late one evening (see also Winkler, 1983b). In spite of their dietary catholicism, foods available to Mono Lake gulls other than brine shrimp appear to be very limited.

Winkler (1983b) states that brine shrimp are not present in "profitable" densities until the chick-rearing period, and early in the season the gulls forage away from the lake (Winkler, 1985; Jehl and Mahoney, 1983). There are, however, no data on what constitutes a "profitable" density and how the overall density of brine shrimp in the lake affects the presence of swarms or patches of higher density. This information is critical for determining the effect of reductions of brine shrimp populations on gull populations. Unlike the situation at Great Salt Lake, with a

major refuse dump within 15 km of the colonies and extensive agricultural fields nearby, Mono Lake has only a few
small refuse dumps nearby and has no nearby agricultural
activity that would sustain a large gull population (Winkler,
1983b).   There is a lack of information on the types and
availability of prey taken by gulls early in the breeding
season, and additional sampling is required to determine the
importance of foods other than brine shrimp throughout the
season.

Summarizing the information on birds, large numbers of
eared grebes, red-necked phalaropes, Wilson's phalaropes,
and California gulls use the lake and depend on the brine
shrimp and brine flies for their food.   Additionally, the
gulls require safe refuges on which to nest.   Present census techniques are statistically inadequate to detect small
but possibly biologically significant changes in avian numbers or temporal patterns of use of Mono Lake.   Likewise,
we do not know critical food densities below which the
invertebrates taken by birds are not sufficient to provide
needed body weight, or how the density of available prey
will change with overall prey densities.   These data are
needed if we are to quantitatively predict how bird populations will respond to changing food supplies with changing
lake salinity.   Clearly, however, if the invertebrate populations are drastically reduced, there will be reductions in
the number of birds visiting and using Mono Lake.

REFERENCES

Blem, C. R.   1980.   The energetics of migration.   Pp. 175-
    224 in Animal Migration, Orientation, and Navigation, S.
    Gauthreaux, Jr., ed. New York: Academic Press.
Bowen, S. T.   1962.   The genetics of *Artemia salina*. I. The
    reproductive cycle. Biol. Bull. 122:25-32.
Bowen, S. T., E. A. Fogarino, K. N. Hitchner, G. L. Dana,
    V. H. S. Chow, M. R. Buoncristiani, and J. R. Carl.
    1985.   Ecological isolation in *Artemia*: population differ-
    ences in tolerance of anion concentrations.   J. Crusta-
    cean Biol. 5:106-129.

Busa, W. B., and J. H. Crowe. 1983. Intracellular *p*H regulates the dormancy and development of brine shrimp (*Artemia salina*) embryos. Science 221:366-368.

Busa, W. B., J. H. Crowe, and G. B. Matson. 1982. Intracellular pH and the metabolic status of dormant and developing *Artemia* embryos. Arch. Biochem. Biophys. 216:711-718.

Chapman, D. J. 1982. Investigations on the salinity tolerance of a diatom and green algae isolated from Mono Lake. Abstract from Mono Lake Symposium, Santa Barbara, Calif., May 5-7, 1982.

Chappell, M. A., D. L. Goldstein, and D. W. Winkler. 1984. Oxygen consumption, evaporative water loss, and temperature regulation of California gull chicks (*Larus californicus*) in a desert rookery. Physiol. Zool. 57:204-214.

Clegg, J. S. 1964. The control of emergence and metabolism by external osmotic pressure and the role of free glycerol in developing cysts of *Artemia salina*. J. Exp. Biol. 41:879-892.

Clegg, J. S. 1974. Interrelationships between water and metabolism in *Artemia salina* cysts: hydration-dehydration from the liquid and vapour phases. J. Exp. Biol. 61:291-308.

Clegg, J. S. 1976a. Hydration measurements on individual *Artemia* cysts. J. Exp. Zool. 198:267-272.

Clegg, J. S. 1976b. Interrelationships between water and cellular metabolism in *Artemia* cysts. II. Carbohydrates. Comp. Biochem. Physiol. 53A:83-87.

Clegg, J. S. 1976c. Interrelationships between water and cellular metabolism in *Artemia* cysts. III. Respiration. Comp. Biochem. Physiol. 53A:89-93.

Clegg, J. S. 1978. Interrelationships between water and cellular metabolism in *Artemia* cysts. VIII. Sorption isotherms and derived thermodynamic quantities. J. Cell. Physiol. 94:123-137.

Clegg, J. S., and F. P. Conte. 1980. A review of the cellular and developmental biology in *Artemia*. Pp. 11-54 in The Brine Shrimp *Artemia*. Vol. 2. Physiology, Biochemistry, Molecular Biology. G. Persoone, P. Sorgeloos, O. A. Roels, and E. Jaspers, eds. Wetteren, Belgium: Universa Press.

Collins, N. C. 1975. Population biology of a brine fly (Diptera: Ephydridae) in the presence of abundant algal food. Ecology 56:1139-1148.

Collins, N. C. 1980a. Population ecology of *Ephydra cinerea* Jones (Diptera: Ephydridae), the only benthic metazoan of the Great Salt Lake, USA. Hydrobiologia 68:99-112.

Collins, N. C. 1980b. Developmental responses to food limitation as indicators of environmental conditions for *Ephydra cinerea* Jones (Diptera). Ecology 61:650-661.

Conover, M. R. 1983. Recent changes in ring-billed and California gull populations in the western United States. Wilson Bull. 95:362-383.

Conover, M. R., and D. O. Conover. 1981. A documented history of ring-billed and California gull colonies in the western United States. Colonial Waterbirds 4:37-43.

Conte, F. P. 1977. Molecular mechanisms in the branchiopod larval salt gland (Crustacea). Pp. 143-159 in Water Relations in Membrane Transport in Plants and Animals, A. M. Jungreis, T. K. Hodges, A. Kleinzeller, and S. G. Schultz, eds. New York: Academic Press.

Conte, F. P. 1984. Structure and function of the crustacean larval salt gland. Pp. 45-106 in Membranes, J. F. Danielli, ed. International Review of Cytology, Vol. 91. Orlando, Fla.: Academic Press.

Conte, F. P., S. R. Hootman, and P. J. Harris. 1972. Neck organ of *Artemia salina* nauplii: a larval salt gland. J. Comp. Physiol. 80:239-246.

Conte, F. P., P. C. Droukas, and R. D. Ewing. 1977. Development of sodium regulation and *de novo* synthesis of Na+K-activated ATPase in the larval brine shrimp *Artemia salina*. J. Exp. Zool. 202:339-361.

Conte, F. P., J. Lowy, J. Carpenter, A. Edwards, R. Smith, and R. D. Ewing. 1980. Aerobic and anaerobic metabolism of *Artemia* nauplii as a function of salinity. Pp. 126-136 in The Brine Shrimp *Artemia*. Vol. 2. Physiology, Biochemistry, Molecular Biology. G. Persoone, P. Sorgeloos, O. A. Roels, and E. Jaspers, eds. Wetteren, Belgium: Universa Press.

Conte, F. P., R. S. Jellison, and G. L. Starrett. In press. Nearshore and pelagic abundances of *Artemia monica* in Mono Lake, Calif. In Saline Lakes, J. M. Melack, ed.

Developments in Hydrobiology. Dordrecht, Netherlands: Dr W. Junk Publishers.

Cooper, S. D., D. W. Winkler, and P. H. Lenz. 1984. The effect of grebe predation on a brine shrimp population. J. Anim. Ecol. 53(1):51-64.

Cramp, S., and K. E. L. Simmons, eds. 1977. Handbook of the Birds of Europe, the Middle East and North Africa, Vol. 1. Oxford, England: Oxford University Press. 722 pp.

Croghan, P. C. 1958. The osmotic and ionic regulation of *Artemia salina*. J. Exp. Biol. 35:219-233.

Dana, G. L. 1981. Comparative Population Ecology of the Brine Shrimp *Artemia*. Master's thesis, San Francisco State University. 125 pp.

Dana, G. L., and P. H. Lenz. 1986. Effects of increasing salinity on an *Artemia* population from Mono Lake, California. Oecologia 68:428-436.

Dana, G. L., C. Foley, G. Starret, W. Perry, and J. M. Melack. In press. *In situ* hatching rates of *Artemia monica* cysts in hypersaline Mono Lake. In Saline Lakes, J. M. Melack, ed. Developments in Hydrobiology. Dordrecht, Netherlands: Dr W. Junk Publishers.

Dick, W. J. A., and M. W. Pienkowski. 1978. Autumn and early winter weights of waders in north-west Africa. Ornis Scand. 10:117-123.

Drinkwater, L. E., and J. H. Crowe. 1986. Physiological Effects of Salinity on Dormancy and Hatching in Mono Lake *Artemia* Cysts. Los Angeles, Calif.: Los Angeles Department of Water and Power.

Drinkwater, L. E., and J. H. Crowe. 1987. Regulation of embryonic diapause in *Artemia*: environmental and physiological signals. J. Exp. Zool. 241:297-307.

Evans, P. R., and D. J. Townshend. In press. Site faithfulness of waders away from the breeding grounds: how individual migration patterns are established. In Proceedings of the 19th International Ornithological Congress, Ottawa.

Hamilton-Galat, K., and D. L. Galat. 1983. Seasonal variation of nutrients, organic carbon, ATP, and microbial standing crops in a vertical profile of Pyramid Lake, Nevada. Hydrobiologia 105:27-43.

114   *The Mono Basin Ecosystem*

Hammer, U. T. 1986. Saline Lake Ecosystems of the World. Monographiae Biologicae 59. Dordrecht, Netherlands: Dr W. Junk Publishers. 616 pp.

Herbst, D. B. 1981. Ecological physiology of the larval brine fly, *Ephydra hians*, an alkaline-salt lake inhabiting Ephydrid. Master's thesis, Oregon State University, Corvallis. 65 pp.

Herbst, D. B. 1986. Comparative Studies of the Population Ecology and Life History Patterns of an Alkaline Salt Lake Insect: *Ephydra (Hydropyrus) hians* Say (Diptera: Ephydridae). Ph.D. dissertation, Oregon State University, Corvallis. 222 pp.

Herbst, D. B. In press. Comparative population ecology of *Ephydra hians* Say (Diptera: Ephydridae) at Mono Lake (California) and Abert Lake (Oregon). In Saline Lakes, J. M. Melack, ed. Developments in Hydrobiology. Dordrecht, Netherlands: Dr W. Junk Publishers.

Iversen, N., R. S. Oremland, and M. J. Kulg. In press. Big Soda Lake (Nevada). 3. Pelagic methanogenesis and anaerobic methane oxidation. Limnol. Oceanogr.

Jehl, J. R., Jr. 1981a. Mono Lake: A vital way station for the Wilson's phalarope. Natl. Geogr. 160:520-525.

Jehl, J. R., Jr. 1981b. Mortality of Waterbirds at Mono Lake, California. Technical Report 81-133. San Diego, Calif.: Hubbs-Sea World Research Institute.

Jehl, J. R., Jr. 1981c. The Biology of Northern Phalaropes at Mono Lake, California, 1981. Technical Report 81-134. San Diego, Calif.: Hubbs-Sea World Research Institute.

Jehl, J. R., Jr. 1982a. Biology of Eared Grebes at Mono Lake, California. Technical Report 82-136. San Diego, Calif.: Hubbs-Sea World Research Institute.

Jehl, J. R., Jr. 1982b. The Biology of Northern Phalaropes at Mono Lake, California, 1982. Technical Report 82-146. San Diego, Calif.: Hubbs-Sea World Research Institute.

Jehl, J. R., Jr. 1983a. Comments on the Annual Census of California Gull Chicks at Mono Lake, California with Special Reference to Populations on the Paoha Islets. Technical Report 83-156. San Diego, Calif.: Hubbs-Sea World Research Institute. 6 pp.

Jehl, J. R., Jr. 1983b. Breeding Success of California Gulls and Caspian Terns on the Paoha Islets, Mono Lake, California, 1983. Technical Report 83-157. San Diego, Calif.: Hubbs-Sea World Research Institute.

Jehl, J. R., Jr. 1983c. The Biology of Eared Grebes at Mono Lake, California. Technical Report 83-136. San Diego, Calif.: Hubbs-Sea World Research Institute.

Jehl, J. R., Jr. 1984. Comments on the Cooperative Gull Census of 1984, with Special Reference to the Paoha Islets. Technical Report 84-167. San Diego, Calif.: Hubbs-Sea World Research Institute.

Jehl, J. R., Jr. 1985. The Cooperative Gull Census of 1985: Comments and Recommendations. Technical Report. 85-183. San Diego, Calif.: Hubbs-Sea World Research Institute. 5 pp.

Jehl, J. R., Jr. 1986. Biology of red-necked phalaropes (*Phalaropus lobatus*) at the western edge of the Great Basin in fall migration. Great Basin Nat. 46:185-197.

Jehl, J. R., Jr., and S. I. Bond. 1983. Mortality of eared grebes in winter of 1982-83. Am. Birds 37:832-835.

Jehl, J. R., Jr., and C. Chase, III. In press. Foraging patterns and prey selection in avian predators: a comparative study in two colonies of California gull. Stud. Avian Biol.

Jehl, J. R., Jr., and D. R. Jehl. 1981. Post-Fledging Mortality of California Gulls. Technical Report 81-135. San Diego, Calif.: Hubbs-Sea World Research Institute.

Jehl, J. R., Jr., and D. R. Jehl. 1982. Post-Fledging Mortality of California Gulls, 1982. Technical Report 82-147. San Diego, Calif.: Hubbs-Sea World Research Institute.

Jehl, J. R., Jr., and S. A. Mahoney. 1983. Possible sexual differences in foraging patterns in California gulls and their implications for studies of feeding ecology. Colonial Waterbirds 6:218-220.

Jehl, J. R., Jr., and S. A. Mahoney. In press. The roles of thermal environment and predation in habitat choice in the California gull (*Larus californicus*). Condor.

Jehl, J. R., Jr., and P. K. Yochem. 1986. Movements of eared grebes indicated by banding recoveries. J. Field Ornithol. 57:208-212.

Jehl, J. R., Jr., D. E. Babb, and D. M. Power. 1984. History of the California gull colony at Mono Lake, California. Colonial Waterbirds 7:94-104.

Jellison, R. 1985. Zooplankton mediated nitrogen and phytoplankton dynamics in a hypersaline lake. In Abstracts of the 48th annual meeting of the American Society of Limnology and Oceanography, June 17-21, 1985, Minneapolis, Minn.

Jellison, R., and J. M. Melack. 1986. Nitrogen supply and primary production in hypersaline Mono Lake. Eos Trans. Am. Geophys. Union 67:974.

Jellison, R., and J. M. Melack. In press. Photosynthetic activity of phytoplankton and its relation to environmental factors in hypersaline Mono Lake, California. In Saline Lakes, J. M. Melack, ed. Developments in Hydrobiology. Dordrecht, Netherlands: Dr W. Junk Publishers.

Kadlec, J. A. 1971. Effects of introducing foxes and raccoons on herring gull colonies. J. Wildl. Manage. 35:625-636.

Kadlec, J. A., and W. H. Drury, Jr. 1969. Loss of bands from adult herring gulls. Bird-Banding 40:216-221.

Kilham, P. 1981. Pelagic bacteria: extreme abundances in African saline lakes. Naturwissenschaften 68:380-381.

Lang, A. R. G. 1967. Osmotic coefficients and water potentials of sodium chloride solutions from 0 to 40°C. Austr. J. Chem. 20:2017-2023.

Lenz, P. H. 1980. Ecology of an alkali-adapted variety of *Artemia* from Mono Lake, California, U.S.A. Pp. 79-96 in The Brine Shrimp *Artemia*. Vol. 3. Ecology, Culturing, Use in Aquaculture, G. Persoone, P. Sorgeloos, O. Roels, and E. Jaspers, eds. Wetteren, Belgium: Universa Press.

Lenz, P. H. 1982. Population Studies on *Artemia* in Mono Lake, California. Ph.D. dissertation, University of California, Santa Barbara. 230 pp.

Lenz, P. H. 1984. Life-history analysis of an *Artemia* population in a changing environment. J. Plankton Res. 6:967-983.

Lenz, P. H., S. D. Cooper, J. M. Melack, and D. W. Winkler. 1986. Spatial and temporal distribution patterns of three trophic levels in a saline lake. J. Plankton Res. 8:1051-1064.

Lovejoy, C., and G. Dana. 1977. Primary producer level. Pp. 42-57 in An Ecological Study of Mono Lake, California, D. W. Winkler, ed. Institute of Ecology Publication No. 12. Davis, Calif.: University of California, Institute of Ecology.

Mahoney, S. A., and J. R. Jehl, Jr. 1984. The Physiology of Migratory Birds on Alkaline Lakes: Wilson's Phalarope and American Avocet. Technical Report 84-172. San Diego, Calif.: Hubbs-Sea World Research Institute. 21 pp.

Mahoney, S. A., and J. R. Jehl. 1985. Avoidance of salt loading by a diving bird at a hypersaline and alkaline lake: eared grebe. Condor 87:389-397.

Mason, D. T. 1966. Density-current plumes. Science 152:354-356.

Mason, D. T. 1967. Limnology of Mono Lake, California. University of California Publications in Zoology No. 83. Berkeley, Calif.: University of California Press.

Melack, J. M. 1983. Large, deep salt lakes: a comparative limnological analysis. Hydrobiologia 105:223-230.

Melack, J. M. 1985. The ecology of Mono Lake, California. Pp. 461-470 in National Geographic Society Research Reports, Vol. 20. 1979 projects. Washington, D.C.: National Geographic Society.

Melack, J. M. In press. Aquatic plants in extreme environments. In Aquatic Vegetation, J. J. Symeons, ed. New York: Elsevier.

Melack, J. M., J. L. Stoddard, and D. R. Dawson. 1982. Acid precipitation and buffer capacity of lakes in the Sierra Nevada, California. Pp. 465-471 in Proceedings of the International Symposium on Hydrometeorology, Denver, Colo., June 13-17, 1982. A. I. Johnson and R. A. Clarke, eds. Bethesda, Md.: American Water Resources Association.

Nichols, W. F. 1938. Some notes from Negit Island, Mono Lake, California. Condor 40:262.

Oremland, R. S., L. Marsh, and D. J. DesMarais. 1982. Methanogenesis in Big Soda Lake, Nevada: an alkaline, moderately hypersaline desert lake. Appl. Environ. Microbiol. 43:462-468.

Oremland, R. S., R. L. Smith, and C. W. Culbertson. 1985. Aspects of the biogeochemistry of Big Soda Lake,

Nevada. Pp. 81-99 in Planetary Ecology, D. E. Caldwell, J. A. Brierley, and C. L. Brierley, eds. New York: Van Nostrand Reinhold.

Palmer, R. S. 1962. Handbook of North American Birds, Vol. 1. New Haven, Ct.: Yale University Press. 567 pp.

Pelagos Corporation. 1987. A Bathymetric and Geologic Survey at Mono Lake, California. Report prepared for Los Angeles Department of Water and Power. San Diego, Calif.

Pienkowski, M. W., and P. R. Evans. 1985. The role of migration in the populations dynamics of birds. Pp. 331-352 in Behavioural Ecology: Ecological Consequences of Adaptive Behaviour, R. M. Sibly and R. H. Smith, eds. British Ecological Society, Vol. 25. Palo Alto, Calif.: Blackwell.

Pitelka, F. D. 1979. Introduction: the Pacific coast shorebird scene. Stud. Avian Biol. (2):1-11.

Power, D. M., P. W. Collins, and K. W. Rindlaub. 1980. The California gull. Pp. 18-286 in The Biology of Certain Water and Shore Birds at Mono Lake, California, D. M. Power and Associates. Unpublished technical report for Los Angeles Department of Water and Power.

Pugusek, B. H., and K. L. Diem. 1983. A multivariate study of the relationship of parental age to reproductive success in California gulls. Ecology 64:829-839.

Rockland, L. B. 1960. Saturated salt solutions for static control of relative humidity between 5° and 40°C. Anal. Chem. 32:1375-1376.

Salzman, A. G. 1982. The selective importance of heat stress in gull nest location. Ecology 63:742-751.

Shuford, W. D. 1985. Reproductive Success and Ecology of California Gulls at Mono Lake, California in 1985, with Special Reference to the Negit Islets: An Overview of Three Years of Research. Contribution No. 318. Stinson Beach, Calif.: Point Reyes Bird Observatory. 50 pp.

Shuford, D., E. Strauss, and R. Hogan. 1984. Population Size and Breeding Success of California Gulls at Mono Lake, California, in 1983. Contribution No. 126. Stinson Beach, Calif.: Point Reyes Bird Observatory.

Shuford, D., P. Super, and S. Johnston. 1985. Population Size and Breeding Success of California Gulls at Mono

Lake, California, in 1984. Contribution No. 294. Stinson Beach, Calif.: Point Reyes Bird Observatory.

Stocco, D. M., P. C. Beers, and A. H. Warner. 1972. Effect of anoxia on nucleotide metabolism in encysted embryos of the brine shrimp. Dev. Biol. 27:479-493.

Storer, R. W., and J. R. Jehl, Jr. 1984. Moult patterns and moult migration in the black-necked grebe *Podiceps nigricollis*. Ornis Scand. 16:253-260.

Thun, M. L., and G. L. Starrett. 1986. The effect of cold, hydrated dormancy and salinity on the hatching of *Artemia* cysts from Mono Lake, Calif. Report to Los Angeles Department of Water and Power.

Wetmore, A. 1925. Food of American Phalaropes, Avocets, and Stilts. U.S. Department of Agriculture Bulletin 1359. Washington, D.C.: U.S. Government Printing Office. 20 pp.

Winkler, D. W., ed. 1977. An Ecological Study of Mono Lake, California. Institute of Ecology Publication No. 12. Davis, Calif.: University of California, Institute of Ecology.

Winkler, D. W. 1979. Deposition in the case: National Audubon Society et al., vs. Department of Water and Power of the City of Los Angeles. California Superior Court, Mono County, Civil No. 6429, Vols. I and II.

Winkler, D. W. 1982. Importance of Great Basin Lakes in Northern California to Nongame Aquatic Birds, 1977. Wildlife Management Branch Administration Report 82-4. Sacramento, Calif.: State of California, The Resources Agency. 30 pp.

Winkler, D. W. 1983a. California Gull Nesting at Mono Lake, California, in 1982: Chick Production and Breeding Biology. Unpublished final report to U.S. Fish and Wildlife Service, Arcata Field Station, Calif.

Winkler, D. W. 1983b. Ecological and Behavioral Determinants of Clutch Size: The California Gull (*Larus californicus*) in the Great Basin. Ph.D. dissertation, University of California, Berkeley. 195 pp.

Winkler, D. W. 1985. Factors determining a clutch size reduction in California gulls (*Larus californicus*): a multi-hypothesis approach. Evolution 39:667-677.

Winkler, D. W., and S. D. Cooper. 1986. The ecology of migrant black-necked grebes at Mono Lake, California. Ibis 128:483-491.

Winston, P. W., and D. H. Bates. 1960. Saturated solutions for the control of humidity in biological research. Ecology 41:232-237.

Young, R. T. 1952. Status of the California gull colony at Mono Lake, California. Condor 54:206-207.

Zink, R. M., and D. W. Winkler. 1983. Genetic and morphological similarity of two California gull populations with different life history traits. Biochem. Syst. Ecol. 11:397-403.

5

Shoreline and Upland Systems

INTRODUCTION

The shoreline and upland systems are integral parts of the Mono Basin. If the level of Mono Lake rises or falls, the shoreline will be inundated or exposed, and the shoreline system will be altered. Many of these alterations in the shoreline system are controlled by hydrologic changes in the nearshore groundwater, as discussed in chapter 2. Except for the streams themselves and the riparian flora and fauna, the upland system will not generally be affected by changes in lake level. A description of the upland system is nevertheless necessary for an overall understanding of the basin.

This chapter describes the physical components of the shoreline and upland systems--topography, soils, and natural events affecting the systems--as well as the biotic components--vegetation and wildlife. The interface between the land and the air, controlling aerosol production from the alkali flats, and the interface between the land and the water, controlling the tufa formations and shoreline erosion, are also discussed.

PHYSICAL COMPONENTS

Topography

The Mono Basin lies on the border of two major physiographic provinces--the Sierra Nevada and the Great Basin--

and is part of both. The first, and still one of the best, descriptions of the Mono Basin was that of I. C. Russell (1889). The basin includes a variety of features of great interest to geologists, climatologists, and geographers--volcanos, fault scarps, glacial cirques and moraines, tufa formations, sand dunes, perennial streams, and several lakes.

The watershed extends to the crest of the Sierra and includes Mt. Lyell, Koip Peak, Mt. Dana, and Mt. Conness. The elevations within the basin range from 13,000 ft to about 6,380 ft, the current level of Mono Lake. The geomorphology of the area is closely related to the geology and the paleoclimatology. Studies of the stratigraphy reveal 12 major glacial advances and various layers of glacial moraines, volcanic pumice and ash, and erosional sediments from streams. Among the terrain features are several of special importance for their scientific interest and scenic value, including Bloody Canyon (a classic example of Sierran glaciation), the Mt. Dana glacier, the Mono Craters, Paoha Island, and Negit Island.

Topographic features in the regions surrounding the Mono Basin include the Inyo Craters, Long Valley, Glass Mountain, the Bodie Hills, Sweetwater Mountain, and the White Mountains (Figure 5.1). Adjoining drainages are the San Joaquin, Tuolumne, and Merced rivers on the western slope of the Sierra, the Walker River to the north, and the Owens River to the south.

Topographic maps and aerial photographs available for the Mono Basin are listed in the bibliography.

Soils

Gallegos (1986) has provided a soils map for the Mono Basin National Forest Scenic Area. He recognized 53 mapping units within the scenic area, with the bulk of the soils belonging to the Entisol order. A few Mollisols and Aridisols were also encountered. Entisols are defined as soils that lack pedogenic horizons except for a slight darkening of the surface layer by organic matter. Mollisols are well-developed soils having a surface layer that is heavily melanized and deep (at least 25 cm deep or one-third of the combined depth of the A- and B-horizons). The

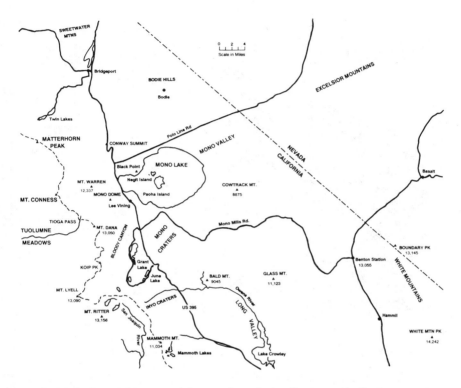

FIGURE 5.1 Topographic and other features of area surrounding Mono Lake.

surface layer has a soft, crumbly structure when dry and
has calcium as the dominant cation on the colloidal
exchange particles. Aridisols have at least one pedogenic
horizon, but never have water continuously available for
plant growth for as much as 90 days when soil temperatures are above 5.0°C.

As Gallegos (1986) has noted, the soils of the scenic
area have developed from two primary parent materials
(Figure 5.2). Soils of areas to the west, southwest, and
northwest of Mono Lake are derived principally from the
granitic core of the Sierra and from metasedimentary rocks
that were uplifted with the Sierra and are now exposed as
scattered fragments along the crests and sides of the
mountain range. These soils are usually coarse textured
and bear variable amounts of rock fragments in the profile.

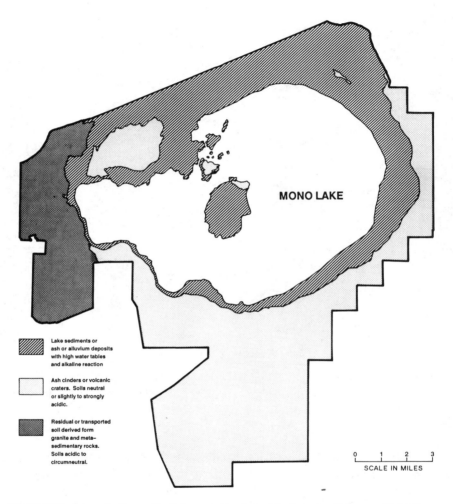

MONO LAKE

Lake sediments or ash or alluvium deposits with high water tables and alkaline reaction	
Ash cinders or volcanic craters. Soils neutral or slightly to strongly acidic.	
Residual or transported soil derived form granite and meta–sedimentary rocks. Soils acidic to circumneutral.	

SCALE IN MILES

FIGURE 5.2 Soils of Mono Basin National Forest Scenic Area.

The rest of the soils of the scenic area are derived from either rhyolitic ash or cinder deposits or from heterogeneous lake sediments. Black Point and Negit Island are both of volcanic origin, but the material is darker (basalt) anc chemically distinct from the rhyolitic Mono Craters. The rhyolitic deposits are young, highly permeable to water, and extremely infertile. The lake sediments are, of course, of mixed origin. Since the lake has progressively receded in its undrained basin, accumulated salts impregnate its younger sediments.

Gentle slopes along the north and east shores of the lake result in large exposures of saline lake sediments and high water tables as the lake recedes. As a consequence, soils along those shores differ strongly from soils of the western and southern shores, where the landscape rises more steeply from the water's edge. In the latter areas, acidic soils occur within a few hundred meters of the highly alkaline, damp shorelines adjacent to the lake. In contrast, sediments that are strongly influenced by the lake with respect to both chemistry and water table often extend for a kilometer or more (sometimes up to 4 km) away from the current shoreline along the north and east shores. Soil salinity problems appear to be exacerbated along these shores by water draining from the Bodie Hills via Wilson Creek. That water becomes highly saline and alkaline as it percolates through the lake sediments. Thus at a large number of sites, water rises to the soil surface by capillarity and leaves behind its load of soluble salts as it evaporates.

The commonest soils on the mountainous west end of the scenic area are Typic Xerorthents (Table 5.1). These are Entisols formed in areas having moist winters and dry summers. The combining term ortho- conveys the idea of genuine or true. Since the suffix -ent designates an Entisol, an orthent soil is a true or genuine Entisol. The extensive morainal deposits in the mountains there support Typic Cryorthents and Typic Cryoborolls. The cryo- prefix designates soils that have a mean annual temperature at 50 cm of over 0° but less than 8.0°C. Rhyolitic outcrops have developed Typic Haploxerolls. Soils ending in -oll are Mollisols or soils with deep, dark-colored epipedons. The moraines and alluvial fans at the mouth of Lundy Canyon support Xeric Torripsamment, Durorthidic Xeric Torripsamment, and Typic Xerorthent soils.

The ash and cinder plains along both east and west sides of the Mono Craters to the south of the lake have developed Dystric Xerorthent, Typic Xeropsamments, Xeric Torripsamments, and Xeric Torriorthent soils (Table 5.1). Dystric soils are dystrophic or infertile due to displacement

TABLE 5.1 The Major Soil Subgroups Encountered on Each of the Three Major Parent Material Types (Figure 5.1) Around Mono Lake

Granite-metasedimentary	Rhyolitic Ash	Lake and Alluvial Sediments
Typic Xerorthents	Xeric Torripsamments	Haplaquents
Typic Cryorthants	Xeric Torriorthents	Durorthidic Xeric
Typic Cryoborolls	Typic Xeropsamments	Torripsamments
Typic Haploxerolls	Dystric Xerorthents	Durorthidic Xeric
Xeric Torripsamments		Torriorthents
Durorthidic Xeric		Aeric Haplaquents
Torripsamments		Typic Psammaquents
Typic Xerorthents		Typic Haplaquents
		Typic Xerorthents
		Xeric Torripsamments
		Xeric Torriorthents
		Xerollic Camborthids

NOTE: Technical names are used for the soil subgroups listed, since they convey information concerning root zone temperature and seasonal water availability, profile development, presence of a water table within the soil profile, and texture of parent material. See Gallegos (1986) for location of the various soils in the landscape.

of biologically essential cations by hydrogen. Such soils are strongly acidic in reaction. Xeric Torripsamments are the most widespread soils in the area, but Xeric Torriorthents and Typic Xeropsamments are also widespread.

The commonest soils on the north and east shores of Mono Lake are Haplaquents, Durorthidic Xeric Torripsamments, Durorthidic Xeric Torriorthents, Aeric Haplaquents, Typic Psammaquents, and Typic Haplaquents (Table 5.1). Aquents are Entisols in which a water table occurs in combination with conditions of poor soil aeration. The prefix hapla- carries the meaning of simple or minimal horizon development. Durorthidic soils have a weakly cemented silicon pan within the surface meter. Black Point supports Xeric Torripsamments and Typic Xerorthent soils. The major upland soil on Paoha Island is a Xeric Torriorthent. The principal upland soil on Negit Island is mapped as a Xerollic Camborthid. Orthids are Aridisols that do not have a high clay or a high sodium horizon. Camborthids have an altered (cambic) subsurface horizon that is generally redder or browner than the surface horizon. These soils are circumneutral to strongly alkaline in reaction. Levels of soluble salts are often so high in some of these soils that all plant life is excluded.

The low fertility of upland soils to the south and west of the lake is striking when parameters for those soils taken in connection with this report are compared with soils from comparable elevations and vegetation types in the Bonneville Basin of western Utah (Table 5.2). The data demonstrate that for most variables, the soils derived from rhyolitic ash contain significantly smaller amounts of elements essential for biological systems than those formed from granitic and metasedimentary parent materials. Both of those Mono Basin parent materials produce soils that are highly impoverished in phosphorus and exchangeable bases relative to the common soils of uplands in the Bonneville Basin (Table 5.2).

Recent experimental plantings of container-grown stock of salt-tolerant native shrubs (*Atriplex canescens* and *Sarcobatus vermiculatus*) on the sandy beaches of the north shore suggest that the erosive action of windblown sand and adverse soil chemistry combine to make revegetation with shrubs an unlikely means of stabilizing such an area (Romney et al., 1986). Direct seeding of grasses and shrubs also shows little promise, but hand plantings of saltgrass (*Distichlis spicata*) rhizomes are often successful on the less harsh portions of the north shore (Romney et al., 1986).

Natural Events

Hydrogeomorphic Events

Hydrogeomorphic events discussed here include avalanches and erosion. The steep eastern escarpment of the Sierra is prone to large, destructive avalanches during and following periods of prolonged snowfall. Particularly affected are the canyons, in which snow sliding downward from the higher slopes is funneled into a narrow valley, thus deepening the mass and increasing its momentum. Avalanches are common in winters when dry, windy periods between major storms create an icy or wind-crusted snow

TABLE 5.2 Some Chemical and Physical Characteristics of Surface (0 to 15 cm) Soils Derived from Rhyolitic Ashes or Granitic and Mixed Metasedimentary Parent Materials in Mono Basin

Location and Parent Material	Sample Size	Textural Class	Organic Matter[a] (%)	pH[b]	Total N[c] (ppm)	Available P[d] (ppm)	Exchangeable[e] (ppm) K	Mg	Ca
Mono Basin rhyolitic ash	18	Loamy sand	2.2 (0.89)	5.1 (0.34)	973.6 (427.26)	5.3 (2.07)	92.7 (34.85)	60.8 (24.93)	769.1 (298.57)
Granite and metasedimentaries	5	Loamy sand	2.1 (0.45)	5.8 (0.42)	1,089.6 (575.02)	9.1 (1.19)	174.0 (49.06)	84.7 (36.72)	1160.2 (331.03)
Bonneville Basin mixed sedimentaries	25	Loam	4.5 (0.83)	7.6 (0.46)	1,367.9 (--)	66.0 (--)	329.0 (144.13)	398.8 (194.41)	7177.5 (4456.57)

[a]Determined by wet digestion using perchloric acid. The data from the Bonneville Basin for organic matter are taken from Harner and Harper (1973) and are based on a sample size of 12.

[b]Determined with a glass electrode on 1:1 soil: distilled water pastes.

[c]Nitrogen was determined by micro-Kjeldahl procedures. Bonneville Basin data are from Ludwig (1969). His sample size was 31, but standard deviations were not reported.

[d]Phosphorus was extracted using 30 ml of 0.2 N acetic acid/g soil. Phosphorus is solution was read colorimetrically using an iron-TCA-molybdate complexing procedure. Bonneville Basin data are from Ludwig (1969).

[e]Exchangeable bases were displaced with 1.0 N ammonium acetate and determined in solution using standard atomic absorption procedures.

NOTE: All sample sites on both parent materials were dominated by sagebrush-bitterbrush (*Artemisia tridentata* and *Purshia tridentata*) vegetation. For comparative purposes, characteristics of soils in the sagebrush-wheatgrass (*A. tridentata* and *Agropyron spicatum*) vegetational zone of the Bonneville Basin in western Utah are presented also. Those data are from Woodward et al. (1984) except as noted. Averages are given for each parameter, with the standard deviation in parentheses below.

128

surface on which the new snow slides easily. Melt-freeze metamorphism is also important in the formation of icy surfaces.

The most widespread and destructive avalanches in recent years occurred throughout the Sierra in February 1986. Damage to vegetation was extensive over a wide range of elevation from mountain hemlock and lodgepole pines near Tioga Pass to aspen and pinyon pines near Mono Lake. Widespread avalanches also occurred in 1921, 1952, 1969, and 1982.

Meteorologic events determine the frequency and severity of erosion episodes along the lakeshore, on lands exposed by the receding lake, and in the steep canyons. As noted below in the section on the land-water interface, strong winds cause lake waves that alter the shoreline and batter nearshore tufa formations. Such winds also blow sand into shifting dunes and transport fine alkali dust, salts, and other particulates to altitudes up to several thousand feet and for distances of tens to hundreds of miles. The extent of wind erosion on surrounding hills and mountain uplands depends in large measure on whether riparian vegetation has been previously destroyed by fire or, less commonly, by overgrazing. Other forms of erosion that affect landforms, particularly streambed and ephemeral stream channel erosion, are associated with high-intensity rainfall and associated flash floods. Such episodes occur most frequently from midsummer to early fall. A particularly memorable event was experienced on Post Office Creek in early August 1955. Heavy thunderstorm rains caused a flash flood that severely damaged Tioga Lodge and washed about 20 automobiles into Mono Lake; no one was killed or injured (H. Klieforth, personal communication).

Fire

In an environment such as that of the Mono Basin, where summer precipitation is scanty and unpredictable, wildfires in the natural vegetation are common. Both low-altitude photographs and space satellite images (Ustin et al., 1986) show conspicuous fire scars in the shrublands and forests of the general area. Woody-stem aging techniques and government records reveal that fires have burned repeatedly throughout at least the past century in the Mono Basin. Historic records demonstrate that the fires result from both natural causes ("dry" lightning) and direct human intervention. Fires are known to have swept over all vegetative types in the basin, including marshes, brushlands, woodlands, and forests.

Because of broken terrain and locally sparse plant cover, few individual fires have burned over large areas. Within the scenic area, there are known scars of over 40 fires that burned in years ranging from before 1875 to 1986, but no fire larger than 100 acres is apparent. Most fires burn fewer than 10 acres before natural factors or direct intervention by fire-control teams limits their spread.

Since plants differ in their ability to regrow after fire, fires in woody vegetation in particular alter the composition of the plant cover for decades after the actual passage of the blaze. Forest trees such as the quaking aspen (*Populus tremuloides*) sprout profusely after fire, while other associated trees (e.g., *Abies concolor* and *Pinus jeffreyi*) are killed by crown fires. At lower elevations or on drier slopes, the major species of the pinyon-juniper woodlands are severely affected by crown fires. *Pinus monophylla*, *Juniperus osteosperma*, *Cercocarpus ledifolius*, and *Artemisia tridentata* all fail to sprout after fire, and soils are stabilized solely by herbaceous species (which are often sparse in these woodlands) for many years after a fire.

In the upland shrublands of the pumice flats south of Mono Lake, neither of the dominant shrubs (*Artemisia tridentata* and *Purshia tridentata*) sprout after fire. As a result, fire scars remain sparsely vegetated for many years on such sites. Grasses and perennial herbs perform poorly on these sites, and adapted annual plants are small and short-lived.

Sprouting shrubs and herbs are the rule on sites nearer the lake, where water tables are near the surface and soils are often at least somewhat saline. The grasses *Distichlis spicata* and *Spartina gracilis* sprout vigorously after fire, as do associated shrubs such as *Chrysothamnus nauseosus*,

Salix exigua, Sarcobatus vermiculatus, and *Shepherdia argentea.*

Since the soils on upland sites in the Mono Basin are generally coarse and well drained, fires on those soils rarely result in erosion by running water. Infiltration rates are rapid enough to preclude the accumulation of surface rivulets that might result in gully formation. On fine-textured sands, fires may permit enough wind action to produce small mounds before natural recovery of the vegetation cover is adequate to prevent soil movement.

Fires on the steeper slopes of the Sierra portion of the scenic area do sometimes result in significant erosion by water. The result may be gentle sheet erosion without the formation of rills or gullies, but topsoil with its content of biologically essential elements does creep slowly downslope. In a few cases, torrential rains or heavy snowpacks have accumulated on fire-denuded slopes and released heavy flows that have produced gullies and moved sediments into stream channels and even into the lake itself. Field observations made while conducting this study suggest that such erosional events in connection with wildfires are not common even along the Sierra front. Of 12 historic wildfire sites examined by a member of the committee, K. T. Harper, only one showed any evidence of significant soil loss from surface runoff. At that site, organic matter in the surface 15 cm of soil was only about 20 percent less than that of adjacent areas that were unaffected by fire. It would thus appear that upland fires produce few erosional events that would significantly affect Mono Lake chemistry directly.

Another adverse effect of fire on steep, wooded slopes along the Sierra face is enhanced frequency of snow avalanches. In some situations, fires appear to have opened avalanche tracks that have not fully healed in a century. Avalanches are not only hazardous to humans, but they also redistribute natural precipitation and alter local runoff of surface water.

While fire is not a serious threat to vegetation near the lake, there are other areas within the Mono Basin where fire could damage the ecosystem. These areas include the lower hills covered by mature sage and bitterbrush stands, the Jeffrey pine forests on the sides of the Mono Craters and other volcanic soils, and the steep slopes west of Mono Lake covered with pinyon pine, juniper, and mountain mahogany. The coniferous forests are vulnerable to particular sequences of events such as long periods of hot, dry weather followed by thunderstorms with much lightning and little rain. Lightning fires followed by gale-force winds can spread ground fires to tree canopies. Such fires are extremely difficult to control in steep terrain. Recovery from such fires may require many years because of slow vegetation growth rates and rapid snowmelt and runoff.

Volcanic Activity and Ashfalls

Ashfalls from volcanic activity have influenced the Mono Basin throughout its history. The oldest known ashfall event occurred about 700,000 years ago. The Aeolian Buttes in the southern portion of the basin are a weathered remnant of that ash deposit. The tan-colored buttes are formed from a welded ash known as Bishop Tuff. This welded ashfall forms a bed some 300 ft thick at a depth of between 1300 and 1600 ft below the surface in a deep well on Paoha Island (Gilbert et al., 1968). The bed appears to be about 500 ft thick in the well at Cain Ranch (Putnam, 1950). The Bishop Tuff eruption was far more massive than any known in historic time and spread a recognizable ash layer over much of western North America. Approximately 125 mi^3 of solids were expelled in the event. For contrast, about 1 mi^3 of material was displaced in the 1980 eruption of the Mount St. Helens in Oregon.

Other major rhyolitic eruptions occurred at various locations over the Long Valley Caldera (from which the Bishop Tuff had been expelled) about 500,000, 300,000, and 100,000 years ago. Each of those events produced ashfalls in the Mono Basin, some 30 mi to the north (Miller et al., 1982). The first of many eruptions along the Mono Crater fracture line occurred approximately 40,000 years ago. About 30 separate domes occur along the north-south axis of the Mono Craters, with the domes becoming progressively younger from south to north. Rhyolitic plugs, cinders, and ash are associated with each eruption. Wood

(1977) suggested that the volume of material ejected per unit time has increased substantially in the last 10,000 years. During the past 2,000 years, eruptions have occurred every 200 to 300 years (Sarna-Wojcicki et al., 1983). The most recent cinder cone along the Mono Craters fracture is Panum Crater, which is only about 625 years old. A more recent upwelling of magma appears to have elevated Paoha Island above water level less than 220 years ago. Although several ash vents spread ash and ejected pumice blocks at the time Paoha Island emerged from the lake, the island itself is composed primarily of lake sediments (S. Stine, University of California, Berkeley, personal communication, 1986).

The Mono Craters and their associated beige-colored ashes are chemically uniform throughout. Black Point, in contrast, is basaltic in nature, as are the deep, black ash deposits that occur around that outcrop (Lajoie, 1968). From 8 to 20 ft of basaltic ash accumulated about 13,300 years ago around what is known today as Black Point. The volcanic plug at Black Point is about 13,500 years old (Lajoie, 1968).

The foregoing volcanic events have lain down dozens of ash layers that lend a distinctively banded aspect to cross sections of sediments underlying Mono Lake. In the Wilson Creek formation alone, 19 separate ash layers occur. The layers range in thickness from a fraction of a millimeter to over 10 cm (Lajoie, 1968). That unique sequence of ash layers has proven to be an invaluable tool for geologists attempting to decipher the history of Mono Lake and its prehistoric antecedent, Lake Russell.

The deep ash layers that occur locally on the national scenic area have left an indelible stamp on contemporary soils and vegetation. The larger prehistoric ashfalls and pumice block ejection events must also have had important if not profound impacts on chemical, physical, and biological characteristics of Mono Lake. As suggested elsewhere in this report, ashfalls may have eliminated fish from prehistoric Mono Basin. The possibility exists that other forms of life were also periodically eliminated by volcanic events in the basin, but events such as these must remain speculative until concrete fossil evidence becomes available. Certainly the record of prior volcanic activity strongly supports the idea that similar events can be expected again in the Mono Basin.

Earthquakes

Strong earthquakes in Mono County occurred on May 25, 1980, and July 20-21, 1986 (Kahle et al., 1986; Kerr, 1985, 1986). Long-range planning for the Mono Basin Scenic Area and other projects in the area cannot ignore the probability of other major earthquakes within the next few decades. The recent seismic activity and related phenomena in Long Valley have been studied by several investigators (Kerr, 1985; Williams, 1985; Linker et al., 1986).

Historic Land Use

Knowledge of early human activities in Mono Basin and archaeological research in the region are summarized and discussed by Aikens (1983). Farquhar (1965) provides an informative account of the history of the Sierra Nevada, while Calhoun (1984) reports on many events of historical and environmental interest in the Mono Basin.

Intensive use of the Mono Basin by nonnative peoples began in the 1850s with activity focusing on explorations for gold and later development of mine properties. Mining activities demanded transportation, lumber, and food. The town of Lee Vining was one of several communities to supply such demands of the mining industry in the 1850s. Lumber mills were built on several of the streams that terminate in Mono Lake. In addition, large flocks of sheep and some cattle were grazed in the area in the late 1800s. Ultimately almost 50,000 acres of farm and associated irrigated pastureland were fenced around the margins of the lake.

Although the level of mining activity in the region around Lee Vining was reduced substantially by the crash of the stock market in 1881, the agricultural industry survived. By 1905, the Southern Sierra Power Company was in existence and acquiring water rights in the Mono Basin for power generation. By 1920 most of the rights to water from Rush and Lee Vining creeks were controlled by either the Southern Sierra Power Company or the Cain Irrigation Company (LADWP, 1984).

In 1923, the City of Los Angeles filed claims on surface waters from several streams tributary to Mono Lake, to supplement waters already being exported from Owens Valley to the south. Claims were filed for surplus waters from Mill, Lee Vining, Walker, Parker, and Rush creeks. In 1930, the citizens of Los Angeles approved a $30 million bond to fund acquisition of water rights on the east slope of the Sierra and construct the Mono Basin-Long Valley water storage facilities. At this time Los Angeles purchased the water rights held by the Southern Sierra Power Company, the Cain Irrigation Company, and several smaller owners. Later in the 1930s, the federal government withdrew all public lands in the Mono Basin from entry in an apparent attempt to protect the water rights held by the City of Los Angeles.

In 1940, California issued permits to Los Angeles to continuously divert up to 200 cubic feet/second (cfs) of water from the Mono Basin and to store 93,540 acre-ft/yr there (LADWP, 1984). This diversion was to be made through the first aqueduct constructed as part of the Mono Basin project. Construction on that aqueduct had begun in 1934 and was completed in 1941. In 1963, Los Angeles began construction of a second aqueduct capable of transporting 210 cfs, with 70 cfs of the water to come from Mono Lake tributaries. The second aqueduct was completed in 1970. In 1984, an average of 138 cfs (about 100,000 acre-ft/yr) were continuously exported from the Mono Basin via the aqueduct system. Since 1974, Los Angeles has held a California state license for diversion of as much as 167,800 acre-ft/yr of water from the Mono Basin (LADWP, 1984).

Los Angeles annually leases about 13,000 acres of unirrigated rangeland in the Mono Basin to private stockmen. An additional 2,000 acres of irrigated pastureland adjacent to Mono Lake are owned by the city and are leased to private operators each year. The city supplies about 8,700 acre-ft/yr of water for irrigation of leased pastures.

Currently the Inyo National Forest manages the federal lands within the Mono Basin National Scenic Area. It issues grazing permits to private operators who graze either sheep or cattle on the public lands of the scenic area. Most of the grazing rights are exercised during late spring to early fall.

Grazing

Cattle have been in the Mono Basin since the 1850s. Large ranches have been major components of the Mono County economy since the early days of settlement; several are still present in the scenic area.

The impacts of grazing by domestic sheep and cattle are probably the most widespread of any related to European people. In the national scenic area the impacts of grazers were concentrated first on the natural wet meadows along the western margins of the lake. As early as 1881, Israel Russell, a pioneer interpreter of the geologic history of Mono Lake, concluded that the wet meadows had been "nearly ruined" by domestic grazers. It is also apparent today that much of the vegetated land between the Pole Line Road and the north shore of the lake has been heavily affected by grazing. Shrub cover has been reduced significantly near areas where fresh water accumulates at the surface, and perennial herbaceous growth is cropped so closely that both escape and nesting cover for wildlife are severely reduced. Annual plants alien to the region now flourish in this portion of the scenic area.

Some (e.g., Davis and Gaines, 1987) have suggested that bunchgrasses and various perennial herbs once filled the interspaces between shrubs throughout the sagebrush zone of the scenic area. This does appear true in the sagebrush zone along the west end of the scenic area, where morainal and alluvial soils are derived from granite and a scattering of metasedimentary parent materials. On the pumice plains to the south of the lake, however, few bunchgrasses or perennial herbs of consequence are apparent in the sagebrush zone unless local conditions result in greater annual precipitation or surface availability of good-quality groundwater. Considering the minimal amounts of calcium and magnesium in the pumice soils and the known inefficiency of grasses in acquiring those ions in competition with dicotyledonous plants (Woodward et al., 1984), it seems likely that bunchgrasses, at least, have never been important on these sites. It is also known that perennial forbs (herbaceous dicots) perform poorly on perennially dry and nutrient-impoverished microsites in the sagebrush zone (Harner and Harper, 1973). Given the low average annual precipitation received by these sites and the nutritionally deficient soils, it is unlikely that perennial forbs ever made a major contribution on the pumice plains.

Roads

The most-traveled paved roads in the Mono Basin are U.S. 395 just west of Mono Lake, Route 120 over Tioga Pass, which passes near the Mono Craters, and the Pole Line Road to Hawthorne. There are many dirt roads in the basin, most of which are used mainly for exploration and recreation. Among the problems that must be confronted by managers of the scenic area is the dust raised by vehicles using these unpaved roads and the increasing invasion of roadless areas by off-road or all-terrain vehicles. Such traffic also disturbs deer and other wildlife. Closing some of the unpaved roads to traffic would probably be in the best interests of both wildlife and soil stability. Certainly such closures would simplify management. Other roads in the area should remain open to the public, but they are vulnerable to drifting sand or snow and would require regular patroling to ensure safety of travelers. Some dirt roads should be paved, and others should be designated for foot travel only.

Mining

From the early days of the settlement of the Mono Basin by nonnative Americans, mining has been a major economic activity. The ghost town of Bodie, now a California state park, was once a temporary home to several thousand miners. Similarly, Aurora, Lundy, and other areas in the region were once active mining complexes. Roads leading to these mountain sites and scars of the mining excavations are visible from Mono Lake and nearby areas. Mining activity is much less prominent now, the principal operation in the scenic area being the pumice excavation at the Mono Craters.   Nevertheless, mining interests remain active.   A particular problem is the habit, not easily controlled, of some mining companies to explore and prospect terrain by bulldozer.   Such actions damage plant, archaeological, and scenic resources.

Logging

Demands for mine timbers, ranch fencing, lumber for homes, and fuel for warmth created another and equally important enterprise, the timber industry.   Hundreds of acres of Jeffrey pines were harvested for lumber, while pinyon pines and junipers were used for fence posts and firewood.   Lumber mills were established on most of the creeks within the Jeffrey pine forests.   Local place names, such as Mill Creek and Mono Mills, are reminders of that history.   Railroads were built to carry the wood to Bridgeport, Bodie, and other towns.   As a consequence of this enterprise, most of the present conifer forest consists of second growth or younger trees.

The timber industry is still active.   In the proposed forest plan for the Inyo National Forest, several large areas are designated for timber harvesting.   The cutting of dead and downed trees for firewood poses a problem for the scenic area.   This practice, often carried out illegally in remote areas where wood-cutting is prohibited, is a threat to wildlife habitat, relict trees, and scenic resources. Curtailment of the use of local wood for camp fires within the scenic area and in adjoining wilderness areas should be considered.

Recreation

The fastest growing component of the economy of the eastern Sierra is recreation.   The town of Mammoth Lakes has grown rapidly during the past 30 years as a result of the popularity of skiing in winter and fishing, hiking, and other vacation activities in summer. Mono Basin is north of the traffic that brings skiers and fishing enthusiasts from southern California to Mammoth Mountain and Crowley Lake, but it is a secondary destination for many of those visitors as well as travelers from Yosemite via Tioga Pass. Plans for the scenic area should include provisions for diverse recreational interests including fishing, boating, hiking, camping, photography, cross-country skiing, and bird and wildlife watching. Some areas should be preserved where such activities could be enjoyed without the intrusion of motorized vehicles.

BIOTIC COMPONENTS

Vegetation

In the following sections, shoreline vegetation and riparian vegetation are discussed separately. In this report, the term riparian vegetation is used to mean vegetation that occurs along streams. Each discussion includes a description of the vegetation and the major environmental factors that influence it. A brief description of the upland vegetation is provided, although this vegetation is unlikely to be affected by changes in lake level.

Shoreline Vegetation

Description of Vegetation Types. Vegetation established on the land exposed as Mono Lake has receded is a patchy mosaic of herbaceous and shrub communities (Figure 5.3). The patchiness has resulted from a complex array of very localized environmental conditions. The controlling variables are primarily soil salinity and available moisture (Bolen, 1964). These variables, in turn, are a consequence of amount of precipitation, water table depth, spring flow, and salt deposition from the lake. The presence of vegetation around springs and areas with shallow water tables is readily seen in a Landsat near-infrared photograph (see back cover).

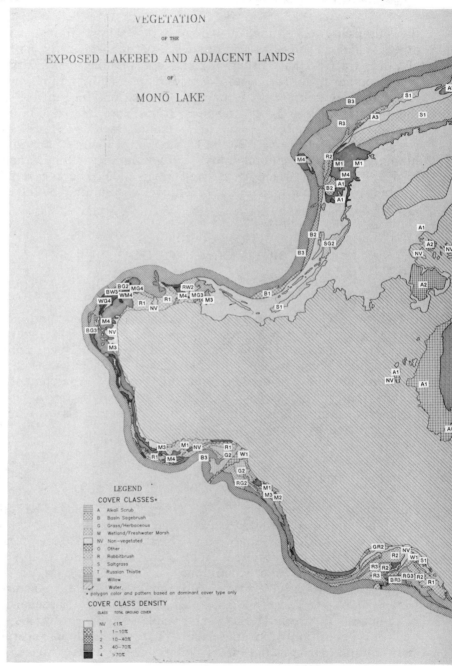

VEGETATION

OF THE

EXPOSED LAKEBED AND ADJACENT LANDS

OF

MONO LAKE

LEGEND

COVER CLASSES*

A   Alkali Scrub
B   Basin Sagebrush
G   Grass/Herbaceous
M   Wetland/Freshwater Marsh
NV Non-vegetated
O   Other
R   Rabbitbrush
S   Saltgrass
T   Russian Thistle
W   Willow
⌁   Water
* polygon color and pattern based on dominant cover type only

COVER CLASS DENSITY

CLASS   TOTAL GROUND COVER

NV   <1%
1   1–10%
2   10–40%
3   40–70%
4   >70%

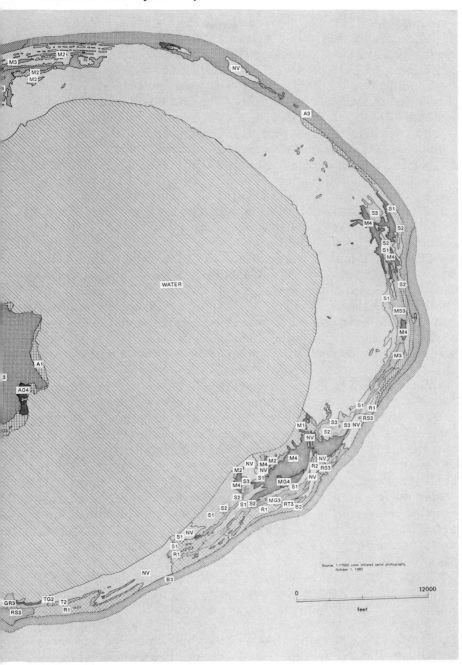

FIGURE 5.3 Vegetation map of the historical shorelands of Mono Lake. Prepared from 1982 aerial photos (Dummer and Colwell, 1985).

There have been few studies of the shoreline vegetation of Mono Lake. In 1982 a vegetation map (Figure 5.3) was made from infared photographs. In 1976 a team of scientists sampled a series of transects selected to represent both typical and unique shoreline phenomena at various locations around the lake using the Braun-Blanquet relevé method (Burch et al., 1977). For example, some transects passed through springs that had been exposed by the receding lake, while others extended across broad expanses of barren, saline exposed lake bed and the vegetated strand and ultimately terminated in an upland shrub community. Except for the rabbitbrush-dominated communities occurring on the raised terraces of the alluvial fans of Lee Vining and Rush creeks, the transects represent most gradients occurring around the lake.

Transect samples were taken in 1976 when lake level was at approximately 6378 ft above sea level. After 1976, the lake dropped to approximately 6372 ft, but more recently it has risen to 6380 ft (November 1986) as unusually heavy runoff has supplied more water than could be transported or stored in water management reservoirs. Rising water has inundated some areas sampled by Burch et al. (1977).

Vorster (1985), in a discussion of the water balance of Mono Lake, analyzed changes in phreatophyte vegetation (plants with roots to the groundwater table) since 1940 and sampled four transects at representative locations along the shoreline of the lake. He showed that between the years 1940 and 1978, phreatophyte communities along the lakeshore increased from 170 acres to 1360 acres. Short-term increases from 1978 to 1982 were difficult to discern from aerial photographs. Phreatophyte communities included any vegetation that used shallow groundwater such as marsh areas with *Scirpus* and *Juncus* as well as dense stands of saltgrass.

Data presented by Burch et al. (1977) and Vorster (1985) from shoreline transects along with information from the vegetation map (Figure 5.3) can be used to create representative cross sections of the topography and vegetation at points along the shoreline. The four cross sections presented in Figure 5.4 are based on locations where groundwater depth and electrical conductivity (EC) were measured in 1986 (see chapter 2). These cross sections do not represent actual vegetation gradients but are representative of existing vegetational gradients that demonstrate the relationship of vegetation types to the lake edge, topography, and groundwater.

At least seven types of associated vegetation occur on Mono Lake shorelines and adjacent upland areas. These types, as used in the cross sections, are described below and are based on personal observations and descriptions, with modifications, from Burch et al. (1977). These vegetational associations correspond well with Vorster's (1985) plant assemblages and assemblages used on the vegetation map (Figure 5.3).

Wet Marsh: Water table at surface from nonlake sources, lush herbaceous growth of *Scirpus americanus*, *Scirpus nevadensis*, *Hordeum jubatum*, *Ranunculus cymbalaria*, *Mimulus guttatus*, *Epilobium adenocaulon*, *Muhlenbergia asperifolia*, *Polypogon* spp., and other species.

Transition Marsh/Dry: Transition areas between marshy and dry environments, soil moisture variable, transition grades either from dry lake sediments upslope into marsh or marsh upslope into drier conditions. Primarily herbaceous growth with variable density from sparse to lush. Herbaceous species include representatives from both marsh and dry habitats.

Alkaline Herbs: Areas with alkaline crusted soils or mud with generally low moisture availability. Vegetation characterized by herbaceous plants such as *Cleomella parviflora*, *Bassia hyssopifolia*, *Puccinellia airoides*, *Scirpus* spp., and *Distichlis spicata* often occurs.

Wet Shrub: Moist areas with a dense herbaceous cover but with more than 5 percent shrub cover. Common herbaceous species include *Scirpus nevadensis*, *Scirpus americanus*, *Distichlis spicata*, and *Juncus effusus*, while shrubs include *Sarcobatus vermiculatus* and *Salix exigua* but not in association.

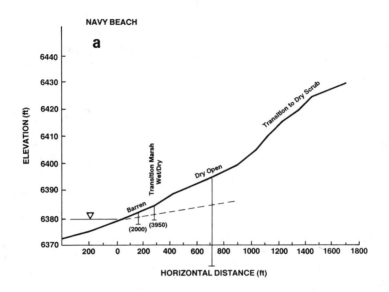

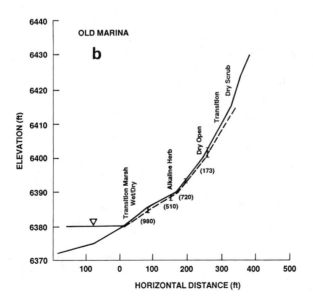

FIGURE 5.4   Profiles of vegetation types relative to shoreline topography, water table, and soil water electrical conductivity (EC) near four representative locations at Mono
Lake.   Lake level (▽) is at elevation 6380 ft.   Ground sur-

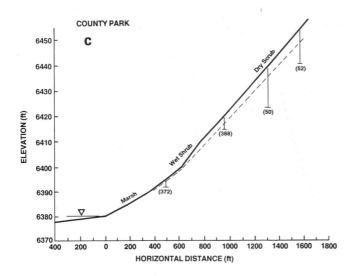

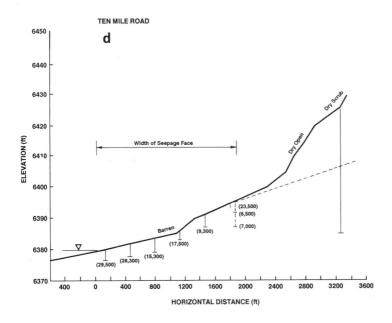

face is a solid line, water table is a dashed line, and soil water EC (μmhos) at piezometers (vertical lines) is given in parentheses. Groundwater data are from fall 1986. Vegetation types are described in the text.

Dry Open: Dry areas with no shrubs and generally very sparse herbaceous cover. Soils are usually well drained and sometimes saline. They typically support *Salsola pestifera, Psathyrotes annua, Bassia hyssopifolia,* and *Distichlis spicata.*

Transition Dry/Shrub: Transition area from dry non-shrub areas into dry shrub areas with shrub cover of 1 to 5 percent.   Herbaceous species are those found in the dry open vegetation type, while shrub species are those found in the dry scrub type.

Dry Scrub: Generally upland vegetation with shrub cover greater than 5 percent and typically a sparse herbaceous layer.   Common shrubs range from *Chrysothamnus* spp. nearer the lake to *Artemisia tridentata, Purshia tridentata,* and *Prunus andersonii* in older (>30 years) stands of upland vegetation.

A brief discussion of each of the four transects both gives a picture of the different gradients from lake to upland around the lake and helps make possible predictions about how shoreline vegetation will change with fluctuations in lake level.

Transect 1 (Figure 5.4a) is located near Navy Beach. Little or no vegetation occurs immediately adjacent to the shoreline. Above this is a marsh area with *Carex, Juncus,* and *Scirpus*.   This grades into a sparse stand of saltgrass (*Distichlis spicata*). Saltgrass increases in density away from the lake.   This saline grassland eventually grades into scattered rabbitbrush (*Chrysothamnus nauseosus*) and greasewood (*Sarcobatus vermiculatus*).   Near the original shoreline of the 1940s, rabbitbrush and big sagebrush (*Artemisia tridentata*) grow in mixed stands to within a few thousand feet of the lake, but rabbitbrush is replaced by bitterbrush (*Purshia tridentata*) in the upland dry scrub association.

Transect 2 (Figure 5.4b) is on the southwest edge of the lake near the Old Marina.   Here marshy vegetation grows near the lakeshore. Bulrush (*Scirpus nevadensis*), *Chenopodium fremontii, Ranunculus cymbalaria,* and foxtail barley (*Hordeum jubatum*) are major constituents of the cover. This vegetation type grades into an alkaline herbs type dominated by saltgrass and smother-weed (*Bassia hyssopifolia*) on its upper edge. Above this area, where the microenvironment is drier, rush and saltgrass are predominant. At still higher elevations, the vegetation is very sparse in the area where the shoreline rises rapidly. There the vegetation is a transition into upland dry scrub.

Transect 3 (Figure 5.4c) is at the northwestern corner of the lake near County Park on a steep shoreline. Vegetation near the lake is wet marsh, dominated primarily by *Scirpus* species. Above this, where the groundwater is still near the surface, a wet shrub community, with arroyo willow as the dominant and rush, muhly grass and bulrush as herbaceous components, occurs. This vegetation grades into an upland dry scrub community with bitterbrush, *Prunus andersonii,* and big sagebrush as dominants.

Transect 4 (Figure 5.4d) is located on the northeastern shore of the lake near Ten Mile Road and crosses an extensive, barren, salt-encrusted shoreline. Except for a few scattered plants, the first 3000 ft of shoreline is barren. The water table in this area is near the surface but is too saline to support vegetation. Above this level near the old shoreline (>6418 ft elevation) the terrain is irregular and the water table becomes deeper. The vegetation closest to the barren shoreline is primarily saltgrass, but rapidly grades into dry scrub characterized by rabbitbrush and big sagebrush. Depending on microenvironmental conditions, the dominant species in the upland area are big sagebrush, rabbitbrush or saltgrass, or a combination of these.

Several general vegetation patterns emerge from analysis of the vegetational cross sections and shoreline studies. For a lake level of approximately 6380 ft, the first 8 to 10 ft above lake level are generally barren. Above this barren area, stands of saltgrass of variable density occur unless the surface is moistened by spring flow or freshwater seepage. If there is a freshwater source, marshy vegetation develops near the lake. Such cover is usually dominated by rush species. Wet areas well removed from the lake often support arroyo willow and are classified as wet shrub vegetation. Edges of the marshy areas are transitional to dry open vegetation, where saltgrass and rush are codominants. Twenty to thirty vertical feet above the shoreline, dry

scrub vegetation appears.   Generally the dry scrub nearer
the lake is characterized by rabbitbrush species.   Above
this, near or above the 1940s shoreline (about 6418 ft), big
sagebrush is initially the characteristic shrub.   In older
stands and in stands on more fertile sites, *Prunus andersonii* or bitterbrush may invade and become codominants
with big sagebrush.

   Successional studies on a beach between Black Point and
Negit Island on black lava sands show similar patterns
(Brotherson and Rushforth, 1985). Above the barren shoreline, three vegetational zones were identified. The zone
closest to the lake was an annual plant zone, the next a
transition zone being invaded by saltgrass, and the third an
area with developed saltgrass vegetation.   Brotherson and
Rushforth showed that saltgrass established at random
points on the beach and then expanded by rhizomes. They
suggested that the vegetational differences were due to
rates of invasion rather than soil differences in this area.
Rather, soil differences may be due to the presence of
saltgrass because saltgrass glands secrete salts and inten-
sify surface salinity of soils where it grows (Hansen et al.,
1976).   The major difference in soil in the three zones was
greater concentration of soluble salts in soils of the saltgrass zone.   The annual plant zone was dominated by
Psathyrotes annua and *Mentzelia torreyi*, while saltgrass
was the primary species in its zone.

Environmental   Factors   Influencing   Shoreline   Vegetation.
Environmental factors influencing the growth and distribution of plants are extensive and varied. However, studies in
the Great Basin have shown that water availability and soil
characteristics may play a more significant role than other
factors (Bolen, 1964).

   Water Availability: The invasion of exposed shoreline
by vegetation is often dependent on the availability of
fresh water from springs or shallow groundwater.   Where
springs occur, marsh vegetation may develop, even near the
lake where soils would normally be barren and salt-
encrusted.   This is demonstrated, for example, at County
Park (Figure 5.4c), where springs occur near the lake, and

at Warm Springs, where they are well removed from the shoreline (see Table 2.2 and Figure 5.3).

Extensive areas of shallow groundwater may also foster development of marshy vegetation, although weak flows may result in sparse cover. This type of marsh development extending across thousands of feet of shoreline occurs at Simon's Spring, which follows a geological fault from the upland out into the lake (see Table 2.2 and Figure 5.3).

If fresh water is not available for vegetation establishment in areas exposed by the receding lake, either the newly exposed shore remains barren, a common situation near current lake edges, or a vegetation generally characterized by saltgrass develops. Saltgrass is usually an indicator of saline conditions rather than of low water availability. As noted above, higher salinity at saltgrass sites may be a result of the plant's ability to exude salts and thus may not be reflective of initial soil conditions (Hansen et al., 1976).

Soils: Soil conditions along vegetational transects show a general gradient from high pH soils (often >9) near the lake to lower pH values on older exposed or upland areas (Burch et al., 1977). Concurrent with this gradient is a general decrease in soil electrical conductivity (EC), an indicator of soluble salts. These patterns do not always hold true, however. In areas with freshwater springs, soil pH may drop below 8.

Areas that may catch windblown soil particles, as on Paoha Island or along the northeastern shoreline (Transect 4, Figure 5.4d), tend to have higher pH and EC values than more sheltered areas such as the western shorelines of the lake.

Burch et al. (1977) reported a general relationship between soil pH and conductivity, and vegetation type. The upland dry scrub vegetation is generally found on soils with the lowest pH and EC observed in the area; dry open, wet marsh, and wet shrub associations have intermediate values; while transition marsh/dry and alkaline herbs associations tend to occur on soils of higher pH and EC. However, local conditions may strongly influence soil chemistry.

Riparian Vegetation

Description of Vegetation Types. Riparian vegetation in the Mono Basin is representative of streamside communities along much of the eastern Sierra Nevada above 1800 m. Taylor (1982) describes nine riparian communities typical of the 1500- to 2800-m elevational zone in the eastern Sierra. Eight of these communities found in the Rush and Lee Vining creek drainages are briefly described below.

Betula occidentalis-Salix lasiolepis: This community of small trees and shrubs constitutes one of the most prevalent of the mid-elevation habitat types of the region. It is best developed between 1200 and 2200 m and is mainly confined to streams with lower flow rates. Mean canopy height is about 8 m. Conifers and tall deciduous trees are rare in this vegetation. A few *Pinus jeffreyi* and *Populus trichocarpa* individuals or clones may be present. This habitat type grades into the *Pinus jeffreyi-Populus trichocarpa* type at its upper elevational limits and into the *Salix laevigata-Elymus triticoides* type at its lower elevational limit. A typical stand of this vegetation may include *Salix rigida* and *S. exigua*, *Rosa woodsii*, *Rhamnus californica*, and herbaceous species of such genera as *Juncus*, *Carex*, *Muhlenbergia*, *Poa*, *Solidago*, *Aster*, and *Mimulus*.

Pinus jeffreyi-Populus trichocarpa: This association creates one of the more important riparian habitat types of the eastern Sierra Nevada. This type is found on most of the larger streams in the region. The two dominant species are usually of about equal importance in the stands. Trees can reach 50 m in height and 2 m in diameter. In addition to the two dominant species, *Populus tremuloides*, *Pinus contorta*, *Abies concolor*, and *Juniperus occidentalis* are frequent. This habitat type grades into *Pinus contorta-Populus tremuloides* vegetation above about 2200-m elevation mostly along low-gradient sections of streams in glacial valleys (e.g., Lee Vining Creek) where soils are well drained. Shrubs that may occur in this vegetation type include species of *Rosa*, *Cornus*, *Sambucus*, *Prunus*, and *Amelanchier*. Herbaceous genera may include *Juncus*,

Agrostis, Carex, Elymus, Poa, Solidago, Epilobium, Aster, and others.

Pinus contorta-Populus trichocarpa: This type is the high-elevation equivalent (above 2200 m) of *Pinus jeffreyi-Populus trichocarpa* vegetation. It lacks some lower-elevation species such as *Betula occidentalis*. Higher-elevation herbaceous species such as *Aquilegia formosa* and *Lupinus pratensis* serve to distinguish this type from the lower-elevation type. This type is typically found along Rush Creek above Grant Lake Reservoir.

Populus tremuloides-Carex lanuginosa: This association is common at mid-elevation along the eastern Sierra streams. Although *P. tremuloides* is not exclusively a riparian species above 2500 m, it is often the riparian dominant. This type, with a canopy of about 15 m, is usually found where there is seepage along the water course. It is one of the most species-rich riparian types along the eastern Sierra. This type can be seen in the glacial valley along Lee Vining Creek at 2200 m.

Carex praegracilis-Juncus balticus: This habitat type occurs in outwash meadows along low grade streams. It is dominated by graminoids and is species-rich. It, like the preceding type, is common along Lee Vining Creek at 2200 m.

Abies concolor-Populus trichocarpa: This association type is infrequent but not unimportant as riparian vegetation. It occurs along steep grade streams above 2200 m. Both dominant species function as facultative riparian species within this elevation zone. *Pinus contorta* may also occur within this type.

Salix rigida-Salix lasiandra: This type is most common along streams above 2200 m with low flow rates. It is more species-rich than its lower elevation counterpart, *Salix exigua-Juncus balticus*. Typical stands occur as narrow bands of willows traversing shrub-steppe desert.

Above 2500 m, riparian plant community types dominate more area. However, at these elevations, riparian species become less obligate and more facultative as riparian and upland habitats become ecologically more similar. These riparian communities are on the outer edges of the Mono Basin and thus have little effect on the lake.

Photographs taken in 1940 have been used to describe the riparian communities of various streams that fed Mono Lake prior to diversion of water by the City of Los Angeles. The two streams most extensively surveyed (Taylor, 1982) were Lee Vining and Rush creeks. Both creeks had extensive riparian communities extending from the glaciated valleys above the points of stream diversion down to the alluvial fans at the lake's edge.

The riparian vegetation along Lee Vining Creek from the area above highways 120 and 395 to an area about halfway between highway 395 and the lake was dominated by an association of quaking aspen (*Populus tremuloides*) and black cottonwood (*P. trichocarpa*). The understory shrubs of this association were *Cornus stolonifera* and *Rosa woodsii*. This vegetation type graded into the Jeffrey pine-black cottonwood (*Pinus jeffreyi-Populus trichocarpa*) (also quaking aspen) habitat type upstream where the creek emerges from the canyon. Quaking aspen occurred in stands from below highway 395 almost to the lake's edge.

Two major associations occurred on the alluvial fan. One was the sandbar willow-wire rush (*Salix exigua-Juncus balticus*) type, which formed thickets. The other was a wire rush-sedge (*Juncus balticus-Carex praegracilis*) association, which formed meadows. The latter of these apparently occurred on poorly drained soils.

Irrigation canals have also developed riparian strand vegetation. These strands include representatives of various riparian communities, ranging from quaking aspen stands to rush meadows.

Photographs taken of Rush Creek below Grant Lake Reservoir in 1940 show riparian vegetation to be as well developed as along Lee Vining Creek. The range of riparian vegetation types was also as variable as on Lee Vining Creek, with extensive stands of black cottonwood and quaking aspen along the channel and willow thickets along the streams just above the alluvial fan.

Photographs taken in 1954 show a great reduction in the riparian communities along both Lee Vining and Rush creeks. Reduction of vegetation along Rush Creek was not as dramatic as along Lee Vining Creek. Flow in Lee Vining Creek above highway 395 was reduced to a small stream, while Rush Creek continued to maintain some riparian vegetation because flows equivalent to about 25 percent of pre-1941 volumes were released during the 1941 to 1954 period. Lateral meadows were desiccated and areas of willow thickets along the channel were greatly reduced because of reduced flows.

Between the highway and the lake, a fire in the early 1950s eliminated most of the woody riparian vegetation along Lee Vining Creek. The quaking aspen strand that had extended almost to the alluvial fan was destroyed. The willow thickets had also largely disappeared and only some rush meadows remained by 1954.

Both Lee Vining and Rush creeks showed accelerated loss of riparian woodlands and buildup of extensive stands of dead or fallen timber during the late 1950s and in the 1960s and 1970s (Taylor, 1982). Periodic releases of water into Rush Creek in the 1970s and 1980s either maintained the remnants of the riparian community or, in some cases, stimulated recruitment and regrowth of riparian species. Thus in 1986, various reaches along Rush Creek showed a revived riparian forest or recruitment of riparian tree or shrub species onto the gravels adjacent to and in the stream channel. *Populus* and *Salix* species appear to have recovered more completely than others under these circumstances.

Taylor (1982) also studied Mill Creek because it had been diverted for hydroelectric power in 1909. Although there was reduced flow along the creek, a well-developed riparian community of quaking aspen remained. An irrigation diversion from Mill Creek built before 1900 supported riparian communities of Jeffrey pine-black cottonwood and sandbar-Pacific willow (*Salix exigua-S. lasiandra*). Below highway 395, there was a declining stand of Jeffrey pine and black cottonwood, while closer to the lake riparian vegetation was mostly gone. Nearer the lake, rising groundwater has maintained a degraded stand of black cottonwood.

Environmental   Factors   Influencing   Riparian   Vegetation.
The major factors influencing riparian vegetation in Mono
Basin are water availability, grazing, and fire.   These fac-
tors are discussed below.

   Water   Availability: Riparian   vegetation   by   definition
occurs on the bank of a body of water, either a river or a
lake.   Thus, availability of water becomes the most influential environmental factor controlling the establishment and
survivability of the community. Water availability is, in
turn, influenced by a variety of other factors, many of
which are associated with valley or stream topography
and/or geology.   These factors include valley type and substrate, varying from the high mountain valleys (either V-
or U-shaped with a bedrock or boulder substrate) to mid-
elevation valleys (usually U-shaped and filled with till down
to the alluvial fans at the foot of the mountains with substrates of gravels or finer fill).   Stream reaches in the
upper valley types tend to be gaining reaches (surface
streamflow increases), while reaches crossing the alluvial
fans tend to be losing reaches (surface streamflow
decreases).   Gaining reaches add water through surface or
subsurface input, while losing reaches lose water through
percolation into the substrate.   Other factors influencing
water availability are stream gradients ranging from steep
at high elevations to shallow on alluvial fans; channel
width (the width of the incised channel along which riparian vegetation establishes); floodplain width (the width of
the portion of the valley that is influenced by periodic
flooding and available groundwater); floodplain cross section (the width of the plain and steepness and height of
the channel and floodplain banks); and streamflow (the
amount of surface water carried by a stream per unit time,
reported as cfs (cubic feet per second) or volumes per year
(e.g., acre-feet per year)).   Subsurface flow is not included
in this term but may be as important.
   The riparian vegetation in the Mono Basin occurs along
streams from the high elevations of the Sierra Nevada
down to the alluvial fans where streams enter the lake.
The high-elevation riparian communities are influenced by
local precipitation, streamflow, and groundwater.   Because
precipitation in the Sierra Nevada is relatively high, many

of the riparian species there function facultatively; that is, they grow both along the streams and in upland communities. These high-elevation streams have steep gradients, little or no alluvial deposit in the channel, and a narrow cross section. Riparian vegetation is limited.

At mid-elevations, for example in the valleys of Lee Vining and Rush creeks about 1000 ft above the lake, the glacial valleys are U-shaped and have sufficient glacial till and alluvium to support extensive riparian stands. The floodplains are relatively wide, and streamflow and groundwater are adequate to maintain any established riparian community. These areas receive so little rainfall in the summer that most riparian species are obligate. The distance over which groundwater spreads from stream channels and saturates the valley fill will influence the width of the riparian strand.

As streams leave the mountain valleys and cross the alluvial fans along the edge of the lake, they lose water to the alluvial deposits. Should lake levels recede, there is enough substrate in such areas to allow streams to downcut through the fill and create deep channels within which riparian vegetation can establish. Temporary or artificial diversions may cause flooding of alluvial fans beyond normal stream channels and permit the establishment of new riparian thickets or meadows.

Most riparian vegetation above Mono Lake but below diversions on the creeks that feed the lake is located on alluvial fans or along shallow-gradient streams that flow through deep channels across the fans. Reaches of streams in these areas are losing reaches. Because the primary source of water for the riparian vegetation in these environments comes from instream flow, flows large enough to balance losses to both evapotranspiration and percolation are required to maintain the vegetation. Inadequate flows will result in desiccation and loss of less drought-tolerant riparian species in such situations.

Most riparian vegetation on the alluvial fans at Mono Lake is currently either depauperate or already dead. Diversion of the streams and resulting low flows across these losing reaches have left insufficient water to maintain a healthy riparian community. In the years since water diversions began in 1941, mortality of riparian species has been greater than recruitment (Taylor, 1982). Thus, riparian vegetation has been reduced or eliminated along most streams below the diversion points.

Recently, streamflow in Rush Creek has been maintained at a sufficient level, either through controlled releases or releases of surplus water, to encourage new recruitment or regrowth of some of the depauperate riparian stands. On the lower part of the alluvial fan, however, the combination of heavy releases, depauperate riparian vegetation, and lower lake levels has caused deep downcutting. Severe channel erosion has undercut and destroyed many remnants of the riparian community and eroded alluvium where recruitment might otherwise have occurred. Locally, along the new channels, some recruitment of riparian species has taken place on floodplain gravels.

The alluvial fan of Lee Vining Creek has not been downcut as much as that of Rush Creek, because there have not been large, long-term releases of surplus runoff. A few riparian plants can be found in the usually dry creek channel, maintained by what must be a limited amount of subsurface flow.

Grazing: Although grazing has had profound effects on riparian systems elsewhere in the semiarid West (Davis, 1982; Kauffman and Krueger, 1984), grazing damage in the riparian areas above Mono Lake cannot be easily demonstrated. Grazing does occur in the upland shrub communities of the Mono Basin. Cattle and sheep would be expected to have access to the basin's riparian vegetation, but two factors reduce the severity of grazing in the riparian zones considered here. First, most of the riparian areas are at the west end of the lake along the streams that rise in the Sierra Nevada. Those streams flow through areas where the land use is more closely supervised than it is in the open rangelands to the east of the lake. Some of the streams also flow through recreational or populated areas, where grazing is controlled. Second, most of the riparian vegetation below the diversion aqueduct has been eliminated or damaged by diversion of much of the flow. Because grazing animals feed on herbaceous understory plants and young woody plants that are depauperate, and because there has been little or no recruitment of woody riparian species, livestock are not attracted to riparian areas below the diversion points. Grazing thus appears to be less significant than water diversions as a factor affecting the vigor of riparian vegetation in the Mono Basin.

Fire: Fire normally is not important in riparian systems because of the moisture in the habitat. However, in the early 1950s a fire along Lee Vining Creek between the diversion and Mono Lake destroyed a large portion of the riparian vegetation, as discussed earlier. Because water flow in the creek was reduced at that time, recruitment of riparian species was limited and full recovery of the riparian community was prevented.

Regeneration and Evapotranspiration of Riparian Vegetation. The two primary riparian genera, *Populus* (cottonwood and aspen) and *Salix* (willow), have not been well studied with respect to the requirements for regeneration along streams (Strahan, 1984; Bradley and Smith, 1985). Most studies demonstrate that both genera establish best in gravelly substrates. *Populus* species usually establish where high streamflows deposit the seeds, in a location that experiences above-normal fluctuations in streamflow and is close enough to the stream to obtain supplemental water. Mature *Populus* does not withstand extensive periods of inundation. Thus, *Populus* does not commonly occur in stream channels but rather on the edge of the floodplains or terraces.

Calculations of evapotranspiration by a water consumptive method for cottonwood (*Populus* spp.) in the western United States vary between 62 and 92 in./yr, with an average 72 in./yr (Blaney, 1961). This early research on water consumption was developed assuming phreatophytes were wasteful and thus the evapotranspiration may be high. However, more recent studies on evapotranspiration, for example Pallardy and Kozlowski (1981) for *Populus* in the Midwest, emphasize individual leaf transpiration and stomatal conductance and are difficult to relate to forest communities. If the 72 in./yr is a high estimate of water consumption by *Populus*, any streamflows established to satisfy this consumption should be more than sufficient for riparian vegetation maintenance. A general average evapotranspiration for willow in the western United States is 54 in./yr (Muckel, 1966). This means that 1 acre of cottonwood will lose 6.0 acre-ft/yr through evapotranspiration and 1 acre of willow will lose 4.5 acre-ft/yr.

Upland Vegetation

Five upland communities were described by Klyver (1931) for the eastern side of the Sierra Nevada. He included alpine communities, subalpine forests, lodgepole pine forests, Jeffrey pine woodlands, and sagebrush. Küchler (1977) mapped and described the following vegetation communities in the Mono Lake area: alpine communities, upper montane and subalpine forests, northern Jeffrey pine forest, juniper-pinyon woodland, and sagebrush steppe.

No truly alpine communities are situated within the Mono Basin National Forest Scenic Area. Likewise, subalpine forests, which are often dominated by *Pinus albicaulis,* do not occur in large enough areas within the scenic area to show up on maps. *Populus tremuloides* is found in more mesic areas of the mountainous portion of the scenic area (Figure 5.5). Jeffrey pine woodlands do occur within the scenic area, both in the west on the Sierra and along the sides of the Mono Craters. This species generally occurs in open, discontinuous stands in the area. It can be found associated with *Juniperus* species, *Purshia tridentata, Arctostaphylos patula, Ceanothus velutinus, Artemisia tridentata* subsp. *vaseyana,* or *Populus tremuloides,* depending on local environmental conditions. At higher elevations, *Abies concolor* and *Pinus contorta* become regular members of older stands.

The juniper-pinyon woodland is usually characterized by *Pinus monophylla* and *Juniperus osteosperma* in the eastern portion of the Mono Basin. *Artemisia tridentata* and *Purshia tridentata* are regular members of this association in the Bodie Hills northeast of Mono Lake and along the foothills of the Sierra Nevada west of the lake. In the latter area *Juniperus osteosperma* drops out of the association. At higher elevations of the Sierra, *Juniperus*

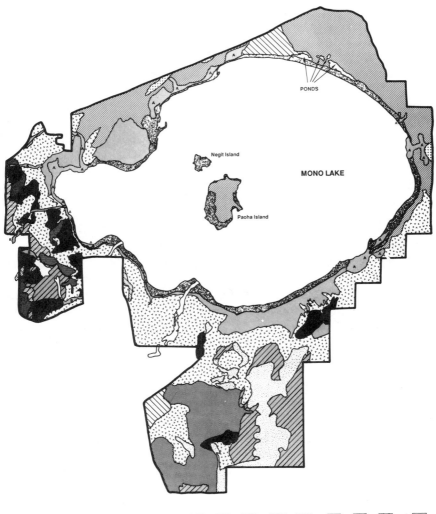

FIGURE 5.5 Vegetation map of Mono Basin National Forest Scenic Area. ARTR = *Artemisia tridentata*, PUTR = *Purshia tridentata*, CHNA = *Chrysothamnus nauseosus*, SAVE = *Sarcobatus vermiculatus*, PRAN = *Prunus andersonii*, GRASS = mixed grasses, CADO = *Carex douglasii*, JUBA = *Juncus balticus*, CHVI = *Chrysothamnus viscidiflorus*, PIJE = *Pinus jeffreyi*.

occidentalis often becomes common, occurring in associations with *Pinus jeffreyi* and other species.

The sagebrush steppe surrounds most of Mono Lake. It is dominated by *Artemisia tridentata* and *Chrysothamnus* species in the early successional stages of the type. Ultimately *Purshia tridentata* becomes codominant with sagebrush on the rhyolitic pumice soils that cover the valley floor south and southeast of the lake (Nord, 1965). *Pinus jeffreyi* may form forested stringers into this association. At lower elevations, *Artemisia tridentata* is represented by subspecies *tridentata*. At higher elevations, as on the shoulders of the Mono Craters and the foothills of the Sierra Nevada, *A. tridentata* subsp. *vaseyana* prevails. The Mono Basin race of *Purshia tridentata* is unusually robust (up to 2 m tall) and upright in form (Nord, 1965).

The sagebrush steppe is far more floristically impoverished on pumice soils than on glacial rubble and alluvial outwash soils derived from the granitic and metasedimentary rocks of the Sierra. On the western margins of Mono Basin, sagebrush steppe vegetation is enriched by a variety of shrubs and subshrubs including *Eriogonum* species, *Leprodactylon pungens*, *Prunus andersonii*, *Tetradymia* species, and *Xanthocephalum*. Associated herbaceous species also contribute much more cover on the western fringes of the basin. Common perennial herbs include *Arabis* species, *Astragalus* species, *Crepis* species, *Oryzopsis hymenoides*, *Phlox longifolia*, *Sitanion hystrix,* and *Stipa* species. Data taken in connection with this study demonstrate that the granitic soils are more fertile than the pumice soils (Table 5.2). Vorster (1985) shows that areas adjacent to the western end of the lake also receive more precipitation than do the pumice flats south of the lake. The combined effects of greater fertility and more precipitation exert a strong influence on composition and productivity of the sagebrush steppe.

Shrubs capable of tolerating both salinity and high water tables become prominent on lake sediments close to the shores of Mono Lake. *Sarcobatus vermiculatus* and *Chrysothamnus nauseosus* are the principle species in such a situation. They dominate extensive areas along the north shore of the lake (Figure 5.5) and are not uncommon along shorelines all around the lake. Saltgrass usually is represented in such communities and forms thick swards where water tables are near the surface. *Artemisia tridentata* subsp. *tridentata* is a common component of these sites where water tables rarely rise higher than 1.0 m below the surface.

At sites where surface flows of water occur seasonally but later disappear completely, *Carex douglasii* and *Juncus balticus* become prominent. Abandoned irrigated fields near the lake often support this association of plants as do many floodplains adjacent to the streams that flow into the lake (Figure 5.5).

Wildlife

The Mono Basin supports a diverse community of terrestrial and aquatic vertebrates, many of which do not rely on the resources of Mono Lake itself. Most of these species are typical of the Sierra Nevada and Great Basin regions and are widespread. For most species, the populations in the Mono Basin represent only small segments of their respective regional populations. A few species, however, deserve special attention because the Mono Basin populations are an important portion of their regional populations, because they are endangered, or because they are the object of recreational hunting or fishing.

Fish

In the geologic past, Mono Lake and its tributaries were connected to the large water system of the Great Basin and shared its fish fauna. As the lake level declined, the basin became isolated from the rest of the system.

When European explorers reached the Mono Basin in the nineteenth century, no fish were found there (Moyle, 1976; Vestal, 1954). This is surprising in view of the geologic and faunal history of the area. The most plausible explanation for the absence of fish is that they were destroyed by volcanic activity; fish bones are present in the streams in the basin beneath volcanic deposits (Moyle, 1976).

Fish have been introduced into the basin since the late nineteenth century, and today the following species are present (Moyle, 1976, and personal communication, University of California, Davis): golden trout (*Salmo aguabonita*), rainbow trout (*S. gairdneri*), cutthroat trout (*S. clarki*), brown trout (*S. trutta*), brook trout (*Salvelinus fontinalis*), Owens sucker (*Catostomus fumeiventris*), tui chub (*Gila bicolor*), mosquitofish (*Gambusia affinis*), threespine stickleback (*Gasterosteus aculeatus*), and Sacramento perch (*Archoplites interruptus*). Of these species, only the cutthroat trout, tui chub, and Owens sucker are native to nearby drainages on the east side of the Sierra; the rest are either west-side species or species native to the eastern United States or Europe. Most of the species are found primarily in lakes in the basin. The principal stream species, in order of abundance, are: brown trout, rainbow trout, brook trout, and threespine stickleback.

At lower elevations, the most important streams for trout fishing are Rush and Lee Vining creeks; in both streams wild brown trout are the most important contributor to the fishery. In Lee Vining Creek, a small, reproducing brown trout population was maintained for 1 to 2 km below the diversion by leakage from the dam and seasonal input from Log Cabin Creek. Summer flows are typically 1 to 2 cfs, but the deep pools in the channel (created at times of higher flows) provide habitat for the adult trout. Prior to the augmentation of flows in 1986, the creek dried up in summer upstream of where it crosses highway 395, causing occasional kills of trout that had colonized the dewatered areas during the winter (P. Moyle, personal communication, 1987).

Rush Creek was completely dry below the diversion until the last several years, when exceptionally heavy snowpacks put more water in the basin than could be diverted. The renewed flows down Rush Creek permitted a large, fast-growing population of brown trout to establish itself (EA Engineering, Science, and Technology, Inc., 1985), and subsequently court-ordered flows of 19 cfs have maintained the fish populations and fishery.

It is difficult to specify minimal flows required to maintain viable populations of trout in lower Lee Vining and lower Rush creeks. It is necessary to have sufficient

continuous flows to provide the habitats needed for reproduction, adult feeding, juvenile feeding, instream production
of aquatic insects, and habitat for juveniles to escape
predation.   Up to some point, a greater streamflow will
support a greater fish population.   In addition, different
species have different requirements.   The Instream Flow
Incremental Methodology (IFIM) model (Bovee, 1982), which
relates streamflow to habitat preference of a species, and
hence is both site-specific and species-specific, is widely
regarded as a reliable method for estimating flow requirements for fish species in streams like Rush and Lee Vining
creeks.   The IFIM model has not yet been applied to those
streams (it has been applied to Lee Vining Creek above the
diversion), but LADWP and the California Department of
Fish and Game plan to remedy this lack in 1987 or 1988.

Birds

Of the over 290 species of birds recorded as visiting or
using the Mono Basin (Gaines, 1986), only four that use the
shoreline or upland habitats are discussed in detail in this
report.   Of these, three species--the bald eagle (*Haliaetus
leucocephalus*), the peregrine falcon (*Falco peregrinus*), and
the snowy plover (*Charadrius alexandrinus*)--deserve special
attention because they are included on state and federal
endangered species lists.   The fourth, the sage grouse
(*Centrocercus urophasianus*), is hunted and may require
special management of its traditional mating-display areas
in the basin.

Bald eagles only pass through the Mono Basin while
migrating, and the small numbers and brief durations of
their visits probably mean that little specific management
action is required on their behalf.   Similarly, peregrine
falcons visit the basin irregularly and in small numbers, but
this might soon change.   Peregrine falcons have been reintroduced into areas near the Mono Basin (Cade and Dague,
1987).   Therefore, if they become established as breeders,
local pairs may use the basin as a hunting area.

Sage grouse, which are hunted, use traditional communal
display areas (leks) during their annual courtship and
mating activities.   While at these leks, whose location in

the basin can be identified, they are sensitive to disturbance, especially of the kind that might occur in the scenic area.

The snowy plover population that nests at Mono Lake was first censused thoroughly in 1978, when 348 breeding adults were detected (Page and Stenzel, 1981). This population accounted for 11 percent of the nesting snowy plovers known in California. No recent censuses of the entire population are available, but intensive work with small segments of the population (e.g., Page et al., 1983) has not suggested any dramatic changes in the size of the population.

The snowy plovers nest primarily on the exposed playa and pumice dunes of the lake's eastern shore. They feed on invertebrates captured at the lakeshore or around springs and seeps. Nests can be located up to 1 km from these feeding sites, and the precocial hatchlings must be led to these areas by their parents in order to feed. Predation of plover chicks by California gulls seems to be the primary mortality factor affecting the population's dynamics. Despite these losses, Page et al. (1983) have found the adult survival and reproductive performance of Mono Lake's snowy plovers to be adequate for intrinsic population maintenance.

The snowy plover population has probably benefited to some extent from recent drops in the lake level because additional areas of playa have been made available for nesting. Because we do not know if the current population is being limited by the available playa area, it is impossible to predict whether or not the population might grow in response to an expansion of the playa by further lake level drops. In any event, there is probably some point beyond which the commuting distance from nest to shoreline feeding areas becomes prohibitively great, so a continued expansion of the width of the playa would not indefinitely expand plover habitat. On the other hand, one can predict with greater certainty that a flooding of the playa area by rising lake levels would be detrimental to the population. A complete inundation of the playa would restrict nesting plovers to the limited area of high pumice dunes. The carrying capacity of these dunes in conjunction with a flooded playa is unknown.

Access to food, rather than space for nesting territories, may limit the snowy plover population. If this limitation pertained, drops in the lake level would result in a concomitant reduction in the length of the shoreline, thus restricting feeding opportunities. A collapse of Mono Lake's invertebrate populations would affect the plover population--as it would other birds that feed on invertebrates--but because plovers also feed around spring-fed seeps, they would probably not totally abandon the Mono Basin as a nesting site.

Because none of the other shoreline or upland bird populations in the Mono Basin constitute such a large portion of their respective regional or continental populations, the committee only summarizes information of their habitat, their seasonal occurrence in the Mono Basin, their status while in the basin, their approximate population size in the basin, and how they are likely to respond to possible disturbances in the basin. Appendix A presents this information for the bird species known to occur in the Mono Basin (Hart and Gaines, 1983); the committee obtained details on each species from a variety of sources, including Grinnell and Miller (1944), Small (1974), Marcot (1979), Hart and Gaines (1983), Gaines (1986), Verner and Boss (1980).

Mammals

Over 70 species of mammals have been reported from the Mono Lake-Tioga Pass region (see Appendix B) (Harris, 1982). Many of these are restricted to the regions of higher elevation and are unlikely to be affected by decisions concerning the management of the scenic area. Others reside in the scenic area, but for reasons of habitat preference or lifestyle are relatively immune to changes in water levels, grazing, or fires. A few species, in particular those that depend on riparian habitats or marshy areas, are sensitive to lake levels and streamflow, while those species that compete with cattle for forage will be affected by grazing pressure and by fires (e.g., white-tailed hare, pygmy rabbit, pronghorn, and bighorn sheep).

In general, there is very little information on the population sizes of any of the resident mammals. A few are known to be abundant or rare, but quantitative data are lacking for all but a very few species. Three species are endemic to the eastern Sierra region: the Inyo shrew, the Panamint kangaroo rat, and the Panamint chipmunk. Pygmy rabbits are endemic to the Great Basin. All occur in the Mono Basin. Three species of shrews found in the Mono Basin are thought to be rare: the Mt. Lyell shrew, the Inyo shrew, and the Merriam's shrew. The first two are probably dependent upon riparian habitats and wetlands. The Merriam's shrew is unusual for a shrew in that it prefers arid habitats, such as the dunes northeast of Mono Lake. The white-tailed hare and the pygmy rabbit are also species that make use of the scenic area and are believed to be rare. Both may be adversely affected by grazing and hunting. Riparian areas may be important for the white-tailed hare. Another species of concern, the mountain beaver, is dependent on access to fresh water and uses riparian areas with dense brush along the west side of Mono Lake (J. Harris, Mills College, personal communication).

Harris (personal communication) details several areas of importance to mammal populations. These areas include riparian and wetland habitats; Black Point, an area with an unusually high diversity of small mammals; and Mono Dunes, with the local race of dark kangaroo mouse (*Microdipodops megacephalus poliontus*), the rare Merriam's shrew, and the Ord's and Panamint kangaroo rats. The dunes are fragile ecological islands, and excessive grazing on these areas may well damage these habitats. The riparian zones provide important habitats for many species and corridors for exchange and movement of mammalian populations between the Sierra and Mono Lake.

LAND-AIR INTERFACE

Environmental Factors Controlling Wind Storms

The interaction between the atmosphere and the land surface has been discussed from different viewpoints in preceding sections of this report. Thus in this section, those discussions will not be repeated. Instead, a few additional aspects are discussed briefly below.

Mountain waves as shown in Figure 5.6 (Holmboe and Klieforth, 1957) are common in advance of approaching Pacific frontal storms. The surface winds are locally strong and are capable of carrying dust, smoke, and other fine particulates as high as 16,000 ft into the atmosphere. They are also capable of transporting sand, salts, other coarse sediments, and loose vegetation such as tumbleweeds and concentrating them in areas where surface winds decrease in speed or converge in direction.

The common diurnal wind pattern in the summer begins with an easterly, upslope breeze in the morning in response to solar heating and convection along the eastern escarpment of the Sierra. During the afternoon, the wind shifts to westerly in response to solar heating of the mountain ranges east of Mono Lake, and probably also to a larger scale diurnal pattern involving the Sierra and the western Great Basin. This diurnal pattern prevails when the synoptic flow aloft (from 10,000 to 20,000 ft above sea level) is weak and from the Pacific. Surface winds associated with this pattern are light to moderate and are not responsible for dust storms or transport of larger particulates.

During episodes of convective cloud development and thunderstorm activity, the diurnal pattern is magnified, leading to the commonly observed sequence shown in Figure 5.7 (from Powell and Klieforth, in press). In this pattern there are strong local downbursts, which frequently cause blowing sand and dust locally and may damage vegetation and structures.

In a recent report, Cahill and Gill (1987) presented the results of a computer model of air quality in the vicinity of Mono Lake in which they predicted the frequency and intensity of particulate levels near Mono Lake for various lake levels and wind velocities. Their model does not appear to consider variations in air mass thermal stability and wind shear, nor does it recognize mesoscale phenomena and diurnal wind patterns. Nevertheless their results are useful for some considerations of air quality.

Another recent report by Saint-Amand et al. (1986) summarized the results of over 10 years of field studies of dust storms from Owens and Mono valleys. They pointed

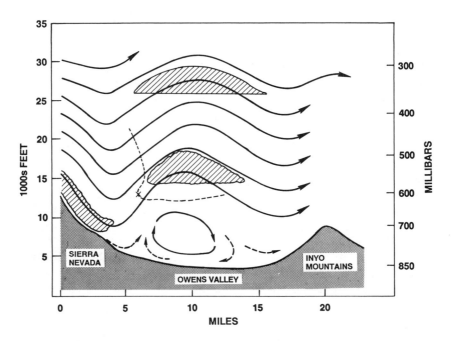

FIGURE 5.6 Vertical cross section of Sierra lee wave showing air flow pattern and cloud forms (Holmboe and Klieforth, 1957).

out that the worst dust storms in Owens Valley were associated with northerly winds aligned with the axis of the valley (parallel to the Sierra), and that such storms transport significant quantities of dust for over 100 mi. In both Owens and Mono basins, strong southerly winds also cause major dust storms. Saint-Amand et al. describe differences between dust episodes in the two basins and discuss possible treatments to alleviate dust problems.

Consequences of Wind Storms

Most of the content of the windborne material is inorganic particulates of geologic origin--sand, salts, and other compounds. These materials when airborne affect visibility and air quality. Larger particulates transported along or

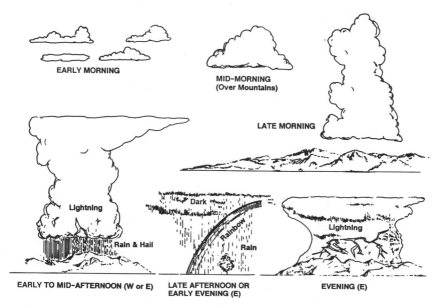

FIGURE 5.7 Typical diurnal sequence of cloud development and precipitation during summer monsoon season (Powell and Kileforth, in press).

near the surface of the ground affect vegetation, tufa formations, wildlife, and human activities. The reports by Cahill and Gill (1987) and Saint-Amand et al. (1986) discuss the composition of airborne particulates and their relation to human respiratory problems. The stabilization of sand dunes, playas, exposed lands, and other erosion-prone terrain has been addressed recently by various groups with consideration of experimental plantings, placement of drift fences, and other treatment. Much more research is needed on all of these problems.

It should be noted that the dust problems of Owens Basin are greater and different in kind from those of the Mono Basin. The town of Lee Vining and nearby population centers are rarely physically affected by airborne dust or blowing sand from the playas surrounding Mono Lake. However, in the future there could be a decrease in air quality caused by smoke and automobile exhausts from heavily populated areas at Mammoth and the June Lake area. Such an increase in aerosols coupled with low-level temperature inversions could also lead to decreased visibility and a possible increase in the frequency and duration of fog over Mono Lake.

LAND-WATER INTERFACE

Tufa Dynamics

The tufa towers, formed when carbonate materials precipitate as described in chapter 3, are a significant scenic attraction of the Mono Basin. As la ke level has declined in the past, groves of lithoid tufa towers have become exposed at the locations and elevations shown in Figure 5.8. These towers range in height from a few feet to tens of feet. The fragile sand tufa, whose locations are shown in Figure 5.9, are castlelike features that form when the carbonate material acts as a cementing agent for sand particles. These formations are not greater than approximately 6 ft in height.

The sand tufa are highly erodible. Wave action associated with changes in lake level could be expected to topple these formations. On the other hand, the lithoid tufa towers are hard and less erodible, although wave action against the base of the towers has been observed to cause towers to topple. Observations at the South Tufa Area by personnel of the Mono Lake Tufa State Reserve suggest that towers that are already unstable may topple with a slow recession of the lake. If the lake level shifts abruptly, otherwise secure towers may be jeopardized. Approximately 24 percent of the changes in tufa formations in the South Tufa Area, one of the most frequently visited tufa areas, appear to have been caused by the wave action from rising lake levels (memo from Dave and Janet Carle, Mono Lake Tufa State Reserve, to Russ Guiney, January 31, 1986).

Shoreline Erosion

As lake levels fluctuate, the shoreline topography will be modified by erosion from wind, surface water runoff, and lakeshore processes. This erosion is significant to the basin ecosystem to the extent that abrasion from wind

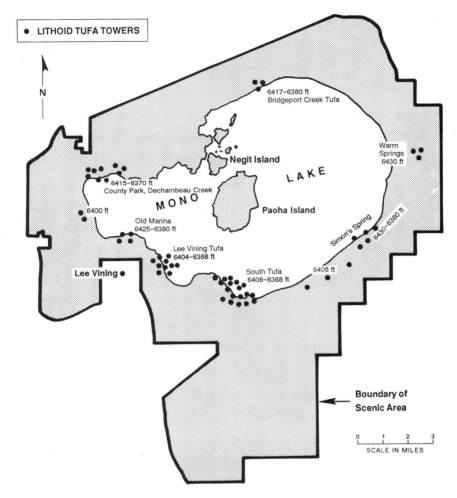

FIGURE 5.8 Locations and elevations of bases of lithoid tufa towers. Elevations are estimated by observations and have not been surveyed. Does not include locations of beach rock or tufa-coated boulders. (Courtesy of N. Upham, U.S. Forest Service.)

inhibits vegetation growth and rapid erosion of surficial soils destroys habitats. No published reports describe these processes in the Mono Basin. Nevertheless, some general observations can be made about the extent of erosion that will occur if lake levels decline.

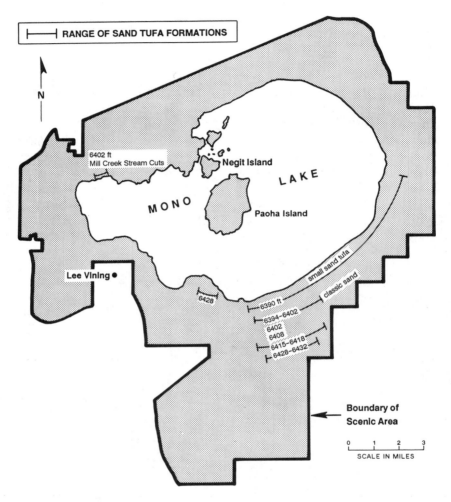

FIGURE 5.9 Locations and elevations of bases of sand tufa. Elevations are estimated by observations and have not been surveyed. (Courtesy of N. Upham, U.S. Forest Service.)

If lake levels drop, several types of shoreline erosion will occur. Declining lake levels will increase the gradient of streams entering the lake, increasing channel erosion in the vicinity of where the streams enter the lake. As the streams adjust to new base levels, the channels will incise, creating steep banks along the channel, increasing sediment load into the lake, and lowering the adjacent water table. The result of the increased sediment load is progression of delta sediments into the lake and perhaps increased turbidity from suspended fine-grained sediments. Stream channel downcutting caused by the past lowering of lake levels is apparent along Rush, Lee Vining, and Mill creeks. In addition, diversion of water from Mill Creek has caused downcutting in Wilson Creek. Even if lake levels remain constant in the future, erosion of bank sediments in these areas, with increased sediment input to the lake, will continue as currently oversteepened banks continue to erode. If lake levels decline in the future, the induced erosion will further incise stream channels, increasing sediment transport to the lake. If lake levels rise, the stream will adjust to a new base level, causing aggradation of channel deposits and decreased sediment load to the lake.

Lowering lake levels would expose large areas of lake bed to erosion by wind, abrasion, surface water, and lake wave action. The effects of wind abrasion are discussed earlier. In addition to downcutting in stream channels by surface water flow, exposed lake beds are subject to rill and sheet erosion by overland flow. This process removes fine sediments from exposed surfaces and increases the transport of sediments to the lake. Prediction of future erosion rates and sediment loads is difficult, however, because the rate of erosion in a particular area depends on a number of factors such as the erodibility of the exposed sediments, the amount of lake level change, and the slope of the exposed lake bottom.

The shoreline process of greatest concern probably is the potential destruction of the islets in the vicinity of Paoha Island (S. Stine, University of California, Berkeley, personal communication, 1987). Geomorphic and erosional features on the islets indicate that they are highly erodible by lake wave action. As with other shoreline erosional processes, however, the extent and rate of erosion that might occur as lake levels change will depend on the amount of lake level change and the number of fluctuations to which the shoreline is subjected.

REFERENCES

Aikens, C. M.   1983.   The Far West.   Pp. 149-201 in Ancient North Americans, J. D. Jennings, ed.   San
Francisco, Calif.: W. H. Freeman.

Blaney, H. F.   1961.   Consumptive use and water waste by
phreatophytes. ASCE J. Irrig. Drain. Div. 87(IR3):37-46.

Bolen, E. G.   1964.   Plant ecology of spring-fed salt
marshes in western Utah. Ecol. Monogr. 34:143-166.

Bovee, K. D.   1982.   A guide to stream habitat analysis
using the Instream Flow Incremental Methodology.   Instream Flow Information Paper 12.   Washington, D.C.:
U.S. Fish and Wildlife Service.

Bradley, C. E., and D. G. Smith.   1985.   Plains cottonwood
recruitment and survival on a prairie meandering river
floodplain, Milk River, southern Alberta and northern
Montana. Can. J. Bot. 64:1433-1442.

Brotherson, J. D., and S. R. Rushforth.   1985.   Invasion and
stabilization of recent beaches by salt grass (*Distichlis
spicata*) at Mono Lake, Mono County, California.   Great
Basin Nat. 45:542-545.

Burch, J. B., J. Robbins, and T. Wainwright.   1977.   Botany.
Pp. 114-142 in An Ecological Study of Mono Lake, California, D. W. Winkler, ed.   Institute of Ecology Publication 12.   Davis, Calif.: University of California, Institute
of Ecology.

Cade, T. J., and P. R. Dague.   1987.   Peregrine Fund Newsletter No. 14.   Ithaca, N.Y.: Cornell Laboratory of
Ornithology.

Cahill, T. A., and T. E. Gill.   1987.   Air Quality at Mono
Lake.   Report to Community and Organization Research
Institute, University of California, Santa Barbara.

Calhoun, M.   1984.   Pioneers of Mono Basin.   Lee Vining,
Calif.: Artemisia Press. 172 pp.

Davis, J. W.   1982.   Livestock vs. riparian habitat
management--there are solutions.   Pp. 175-184   in
Proceedings of the Wildlife-Livestock Relationships Symposium, April 20-22, 1981, Coeur d'Alene, Idaho, J. M.
Peek and P. D. Dalke, eds.   Moscow, Idaho: University
of Idaho Forest, Wildlife and Range Experiment Station.

Davis, L., and D. Gaines.   1987.   Grazing away the scenic
area: report looks at pastoral problems.   Mono Lake
Newsletter 9(3):13-14.

Dummer, K., and R. Colwell. 1985. Vegetation map of the Historical Shorelands of Mono Lake, California. Scale 1:24,000. Prepared from 1982 aerial photos. Prepared by the State of California for *State of California* vs. *U.S.* Civil No. S-80-696. U.S.D.C.E.D. Calif.

EA Engineering, Science, and Technology, Inc. 1985. Trout Population Size and Habitat Characteristics of Rush Creek Between Grant Lake Dam and Mono Lake. Prepared for the Los Angeles Department of Water and Power. 6 pp.

Farquhar, F. P. 1965. History of the Sierra Nevada. Berkeley, Calif.: University of California Press. 262 pp.

Gaines, D. 1986. Birds of Yosemite and the East Slope. Lee Vining, Calif.: Artemisia Press.

Gallegos, J. 1986. Mono Basin National Forest Scenic Area soils analysis. Unpublished report to the Inyo National Forest, Bishop, Calif.

Gilbert, C. M., M. N. Christensen, Y. Al-Rawi, and K. R. Lajoie. 1968. Structural and volcanic history of Mono Basin, California-Nevada. Pp. 275-329 in Studies in Volcanology: A Memoir in Honor of Howel Williams. Geol. Soc. Am. Mem. 116. Washington, D.C.: U.S. Geological Survey.

Grinnell, J., and A. H. Miller. 1944. The Distribution of the Birds of California. Pacific Coast Avifauna No. 22.

Hansen, D. J., P. Dayanandan, P. B. Kaufman, and J. D. Brotherson. 1976. Ecological adaptations of salt marsh grass *Distichlis spicata* (Gramineae), and environmental factors affecting its growth and distribution. Am. J. Bot. 63:635-650.

Harner, R. F., and K. T. Harper. 1973. Mineral composition of grassland species of the Eastern Great Basin in relation to stand productivity. Can. J. Bot. 51:2037-2046.

Harris, J. H. 1982. Mammals of the Mono Lake-Tioga Pass Region. Lee Vining, Calif.: Kutsavi Books. 55 pp.

Hart, T., and D. Gaines. 1983. Field Checklist of the Birds of the Mono Basin. Lee Vining, Calif.: Mono Lake Committee.

Holmboe, J., and H. Klieforth. 1957. The effect of the Sierra Nevada on a Pacific storm. Pp. 99-117 in Investigations of Mountain Lee Waves and the Air Flow over the Sierra Nevada. Final Report to the U.S. Air

Force on Contract AF 19(604)-728. Los Angeles, Calif.: University of California, Meteorology Department.

Kahle, J. E., W. A. Bryant, and E. W. Hart. 1986. Fault rupture associated with the July 21, 1986 Chalfant Valley earthquake, Mono and Inyo counties, California. Calif. Geol. 39(11):243-245.

Kauffman, J. B., and W. C. Krueger. 1984. Livestock impacts on riparian ecosystems and streamside management implications: a review. J. Range Manage. 37(5):430-438.

Kerr, R. A. 1985. Inyo Domes drilling hits pay dirt. Science 227:504-505.

Kerr, R. A. 1986. Do California quakes portend a large one? Science 233:1039-1040.

Klyver, F. D. 1931. Major plant communities in a transect of the Sierra Nevada Mountains of California. Ecology 12:1-17.

Küchler, A. W. 1977. The map of the natural vegetation of California. Pp. 909-938 (plus map) in Terrestrial Vegetation of California, M. G. Barbour and J. Major, eds. New York: Wiley.

Lajoie, K. R. 1968. Late Quaternary Stratigraphy and Geologic History of Mono Basin, Eastern California. Ph.D. dissertation, University of California, Berkeley. 379 pp.

Linker, M. F., J. O. Langbein, and A. McGarr. 1986. Decrease in deformation rate observed by two-color laser ranging in Long Valley Caldera. Science 232:213-216.

Los Angeles Department of Water and Power. 1984. Background Report on Geology and Hydrology of Mono Basin. Report of the Aqueduct Division, Hydrology Section. Los Angeles, Calif.

Ludwig, J. A. 1969. Environmental Interpretation of Foothill Grassland Communities of Northern Utah. Ph.D. dissertation, University of Utah, Salt Lake City. 100 pp.

Marcot, B. G., ed. 1979. California Wildlife/Habitat Relationships Program. Eureka, Calif.: U.S. Forest Service, Six Rivers National Forest. 899 pp.

Miller, C. D., D. R. Mullineaux, D. R. Crandell, and R. A. Bailey. 1982. Potential Hazards from Future Volcanic Eruptions in the Long Valley-Mono Lake Area, East-central California and Southwest Nevada: A Preliminary Assessment. U.S. Geological Survey Circular 877. Washington, D.C.: U.S. Geological Survey. 10 pp.

Moyle, P. B. 1976. Inland Fishes of California. Berkeley, Calif.: University of California Press. 405 pp.

Muckel, D. C. 1966. Phreatophytes--water use and potential water savings. ASCE J. Irrig. Drain. Div. 92(IR4):27-34.

Nord, E. C. 1965. Autecology of bitterbrush in California. Ecol. Monogr. 35:307-334.

Page, G. W., and L. E. Stenzel. 1981. The breeding status of the snowy plover in California. West. Birds 12:1-40.

Page, G. W., L. E. Stenzel, D. W. Winkler, and C. W. Swarth. 1983. Spacing out at Mono Lake: breeding success, nest density, and predation in the snowy plover. Auk 100:13-24.

Pallardy, S. G., and T. T. Kozlowski. 1981. Water relations of *Populus* clones. Ecology 62:159-169.

Powell, D., and H. Klieforth. In press. Water and climate. In The Natural History of the White-Inyo Range, C. Hall, ed. Los Angeles, Calif.: University of California Press.

Putnam, W. C. 1950. Moraine and shoreline relationships at Mono Lake, California. Geol. Soc. Am. Bull. 61(2):115-122.

Romney, E. M., R. B. Hunter, S. D. Smith, and D. P. Groeneveld. 1986. Tests to determine survival of transplanted shrub and grass vegetation on the dust-source shoreland of Mono Lake. Draft final report to Environmental Monitoring and Services, Inc. Los Angeles, Calif.: University of California, Laboratory of Biomedical and Environmental Sciences.

Russell, I. C. 1889. Quaternary history of the Mono Valley, California. Pp. 261-394 in U.S. Geological Survey Eighth Annual Report. Washington, D.C.: U.S. Geological Survey.

Saint-Amand, P., L. A. Mathews, C. Gaines, and R. Reinking. 1986. Dust Storms from Owens and Mono Valleys, California. NWC TP6731. China Lake, Calif.: Naval Weapons Center. 79 pp.

Sarna-Wojcicki, A. M., D. E. Champion, and J. O. Davis. 1983. Holocene volcanism in the conterminous United

States and the role of Silicic volcanic ash layers in correlation of latest-Pleistocene and Holocene deposits. Pp. 52-57 in Late-Quaternary Environments of the United States, Vol. 2: The Holocene, H. E. Wright, Jr., ed. Minneapolis, Minn.: University of Minnesota Press.

Small, A. 1974. The Birds of California. New York: Winchester Press. 310 pp.

Strahan, J. 1984. Regeneration of riparian forests of the Central Valley. Pp. 58-67 in California Riparian Systems: Ecology, Conservation, and Productive Management, R. E. Warner and K. M. Hendrix, eds. Berkeley, Calif.: University of California Press.

Taylor, D. W. 1982. Riparian Vegetation of the Eastern Sierra: Ecological Effects of Stream Diversions. Contribution No. 6. Lee Vining, Calif.: Mono Basin Research Group.

Ustin, S. L., J. B. Adams, C. D. Elvidge, M. Rejmanek, B. N. Rock, M. O. Smith, R. W. Thomas, and R. A. Woodward. 1986. Thematic mapper studies of semiarid shrub communities. Bioscience 36:446-452.

Verner, J., and A. S. Boss. 1980. California Wildlife and Their Habitats: Western Sierra Nevada. U.S.D.A. Forest Service General Technical Report PSW-37. Berkeley, Calif.: U.S. Forest Service, Pacific Southwest Forest and Range Experiment Station. 439 pp.

Vestal, E. H. 1954. Creel returns from Rush Creek Test Stream, Mono County, California, 1947-1951. Calif. Fish Game 40:89-104.

Vorster, P. 1985. A Water Balance Forecast Model for Mono Lake, California. Master's thesis, California State University, Hayward. Earth Resources Monograph No. 10. San Francisco, Calif.: U.S. Forest Service, Region 5.

Williams, S. N. 1985. Soil radon and elemental mercury distribution and relation to magmatic resurgence at Long Valley Caldera. Science 229:551-553.

Wood, S. H. 1977. Distribution, correlation, and radiocarbon dating of late Holocene tephra, Mono and Inyo craters, eastern California. Geol. Soc. Am. Bull. 88:89-95.

Woodward, R. A., K. T. Harper, and A. R. Tiedemann. 1984. An ecological consideration of the significance of cation-exchange capacity of roots of some Utah range plants. Plant Soil 79:169-180.

6
Ecological Responses
to Changes in Lake Level

INTRODUCTION

Changes in the level of Mono Lake from the current
level of 6380 ft can be expected to affect the Mono Basin
ecosystem in various ways. The aquatic biota, birds, and
other wildlife, scenic tufa formations, shoreline and riparian vegetation, lake water chemistry, and air quality will
all be affected in some way by a rising or falling of the
lake level.

Because these aspects of the basin can be expected to
respond differently to changes in lake level, the committee
assessed the consequences to each of these components for
a range of lake levels. In this chapter, the ecological responses to changes in lake level are assessed for levels in
10-ft intervals from 6430 ft, the approximate historic high
stand, to 6330 ft, the approximate stabilization level--i.e.,
the level at which inflows equal outflows and equilibrium
occurs--assuming exports of 100,000 acre-ft/yr and climatic
conditions similar to those in the previous 40 years
(LADWP, 1987; Vorster, 1985). If the climate were to become drier, the stabilization level would be lower. The
ecological consequences of these lower levels were not
considered because the effects would be much the same as
those that would occur if the lake level fell to 6330 ft.

Figure 6.1 illustrates the appearance of the lake and
shoreline at representative lake elevations of 6400, 6380
(current lake level), 6360, and 6340 ft above sea level. Figure 6.2 is a composite representation of the shoreline at

179

The Mono Basin Ecosystem

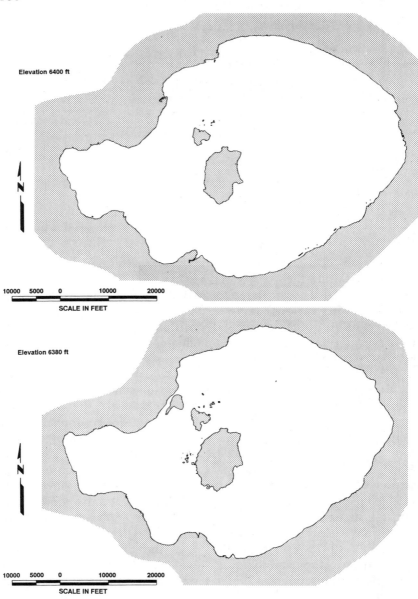

FIGURE 6.1 Shoreline of Mono Lake at lake elevations of 6400 and 6380 ft above sea level.

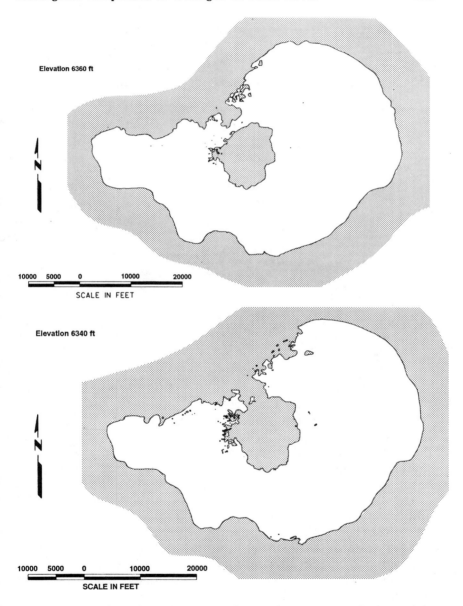

Elevation 6360 ft

10000 5000 0 10000 20000
SCALE IN FEET

Elevation 6340 ft

10000 5000 0 10000 20000
SCALE IN FEET

FIGURE 6.1 (continued) Shoreline of Mono Lake at lake elevations of 6360 and 6340 ft above sea level.

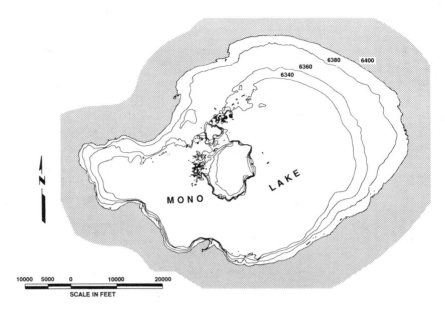

FIGURE 6.2 Composite representation of shoreline of Mono Lake at elevations of 6400, 6380, 6360, and 6340 ft above sea level.

these elevations. If the lake level were to rise above or fall below the current level, the shoreline would be either inundated or exposed. These changes have implications for the snowy plover that use the alkali flats for nesting, the shoreline vegetation, the air quality in and around the basin, and the tufa formations. If the lake level were to drop, the islands would become peninsulas, affecting the ability of California gulls to use the islands for nesting. At lower lake levels the gently sloping incline, particularly on the eastern and northern portions of the lake, would be exposed and the nearshore region of the lake bottom would be more steeply inclined. This result of the bathymetry decreases the availability of the shallow littoral habitat necessary for the maintenance of the brine fly population. Most importantly, the bird populations would be unable to rely on Mono Lake if the lake level decreased to the point where the concomitant increase in salinity eliminated their food source, the brine shrimp and brine fly. Changes in lake level are also associated with changes in the flow of the streams feeding the lake, with consequences for stream biota, the riparian vegetation, and the wildlife populations relying on the riparian habitat.

Lake level is controlled not only by the amount of water exported from the basin, but also by natural climatic conditions. Regardless of the lake level that may be established as necessary to preserve the ecosystem, fluctuations in lake level will occur because of natural changes in weather from year to year. Therefore, predictions about the ecological consequences that would occur at a particular lake level must take into account the possibility of fluctuations around that level.

The effects of changes in lake level on the individual components of the ecosystem--salinity and chemical stratification of the lake, aquatic biology, bird populations, shoreline environment, and upland environment--are discussed below. Later in the chapter, the overall ecological consequences of changes in lake level on the resources of the Mono Basin are summarized.

RESPONSES OF ECOSYSTEM COMPONENTS TO CHANGES IN LAKE LEVEL

Salinity and Chemical Stratification of Lake Water

Because there are no outlets from Mono Lake, ions dissolved in the streamflow and groundwater that feed the lake become concentrated as evaporation causes the lake level to fall. If the lake level rises, the concentration of ions would decrease with dilution from the incoming fresh water.

Using both historical measurements and calculated data from hydrological models, one can estimate the salinity of Mono Lake at specific lake levels, as described in chapter 2. Historical measurements of lake level and salinity are presented in Table 6.1. Predicted values for the salinity of the lake at levels below the current lake level (approximately 6380 ft) are given in Table 6.2. These values were calculated by using the LADWP hydrologic model and the

TABLE 6.1 Historical Measurements of Lake Elevation and Salinity for Mono Lake (LADWP, 1986)

Lake Elevation (ft)	Salinity (g/l total dissolved solids)
6417	51.3
6414	54.0
6410	56.3
6407	58.1
6403	60.2
6380	89.3
6378	86.8
6377	91.6
6376	89.3
6375	93.4
6373	97.7
6372	99.4

TABLE 6.2 Predicted Salinity Values for Lake Elevations Below the Present-day Level of Mono Lake (LADWP, 1986)

Lake Elevation (ft)	Salinity (g/l total dissolved solids)
6370	101.6
6365	110.9
6360	121.4
6355	133.6
6350	147.8
6345	164.6
6340	184.5
6335	208.2
6330	237.2
6325	273.1

area capacity information from the recent bathymetric survey of the lake (Pelagos Corporation, 1987).

The predicted values in Table 6.2 should be treated as upper bounds for the salinity because the LADWP hydrologic model assumes that mineral precipitation does not occur over the range of lake levels modeled. However, geochemical models (R. J. Spencer, University of Calgary, personal communication, 1986) indicate that minerals such as trona, mirabilite, and natron will begin to precipitate at low temperatures (near 0°C) once the salinity of the lake exceeds approximately 125 g/l. Therefore, surface water salinities should be less than the models predict at lake levels below the level at which these minerals begin to precipitate. This occurs because minerals precipitate throughout the water column at low temperatures in the winter and then redissolve near the bottom at higher temperatures in the summer. Consequently, a highly saline layer may develop at the bottom of the lake that is supplied by salts precipitated from the upper regions.

In addition to the chemical stratification caused by precipitation of minerals, a less intense stratification occurs with large inputs of fresh water. The less dense fresh water does not completely mix with the denser saline lake water, creating a condition known as meromixis. The observation that Mono Lake became meromictic following the large freshwater inflows in the spring of 1983 raises the possibility that periods of meromixis have occurred in the past and can occur in the future. Historical records of snow pack sizes indicate that years with large runoff occurred in 1862, 1890, 1938, 1952, 1967, and 1969, as well as in 1983 and 1986. The probability of meromixis depends upon the relative volumes of lake water and inflows and on the density difference between the two waters. Therefore, at higher stands when the lake is less saline and larger in volume, the likelihood of meromixis would be reduced. In contrast, as level and volume decline and salinity increases, the probability that a large runoff will cause persistent chemical stratification increases.

Measurements of limnological and biological conditions in the years following the onset of meromixis in 1983 (chapters 3 and 4) can be used to generalize the implications to the aquatic ecosystem of periods of meromixis.

The persistent and strong stratification in the meromictic lake reduces rates of vertical mixing and results in increasing concentrations of reduced compounds such as methane, ammonium, and hydrogen sulfide in the water below the chemocline. The vertical flux of substances, such as ammonium, across the chemocline depends upon the coefficient of eddy conductivity, which is reduced by the strong density gradient, and upon the concentration gradient of the substance. Hence, while reduced vertical flux of ammonium was observed in 1984-1985 in the initial stages of meromixis, if the density gradient weakens and the concentration gradient of ammonium increases, the vertical flux may increase. Furthermore, if entrainment of water within the chemocline or complete turnover occurred, a large pulse of anoxic water rich in reduced compounds would mix into the upper water. In contrast, the formation of a highly saline layer near the bottom of the lake as a result of the settling and redissolution of minerals should cause the lake to remain meromictic for a very long time. This dense, anoxic layer would trap nutrients within it.

Several biological responses to meromixis are possible. Reduced supply of ammonium will reduce phytoplankton growth and abundance in the upper mixed layer. However, it is possible that photoautotrophic and chemoautotrophic bacteria may grow in the chemocline and saline bottom waters and augment the lake's primary productivity, though probably not sufficiently to replace the loss in algal growth. With lower algal abundance the food available to the first generation of brine shrimp would be reduced, the fecundity of the females would decline, and the switch to cyst production from live bearing could occur earlier each year. A smaller second generation of brine shrimp would result. If a sudden mixing event that injected high concentrations of reduced substances toxic to brine shrimp occurred, an abrupt, major decline in brine shrimp abundance would be likely. Such an event would also add ammonium, which would enhance algal growth.

Another consequence of meromixis is a more dilute upper mixed layer. Hence, if the salinity of the lake had reached a concentration that was adversely affecting the biota, the freshening of the surface waters could improve conditions.

Aquatic Biology

Falling lake levels in Mono Lake would result in increasing salinity with concomitant effects on the aquatic biota. A summary of the impacts on planktonic and benthic algae is provided in Table 6.3 and on brine fly and brine shrimp populations in Table 6.4. Biological responses to meromixis, which would become more likely if lake levels fell, might have synergistic effects with those related to salinity per se.

A gradual decrease in phytoplankton productivity would be expected as the salinity increased to approximately 150 g/l. A greater decrease in growth at salinities above approximately 175 g/l would occur if the current species retained dominance. If more salt-tolerant genera such as *Dunaliella* attained dominance, which is likely, a lesser decrease in phytoplankton productivity at salinities greater than 150 g/l would be probable. One major component of the benthic alga assemblage, *Ctenocladus circinnatus*, would become less abundant and less productive at a salinity of approximately 100 g/l. A much greater decline in overall phytobenthic productivity is expected at approximately 150 g/l.

If lake levels were to fall and expose highly erodible sediments, the combination of upland runoff and wave action could transport these sediments into the lake. The potential influence of these particles on the plankton is complex (Melack, 1985). Increased suspended sediments would decrease transparency, which would decrease the region where primary productivity can occur and may retard feeding by predators such as grebes, who use vision to locate prey, and filter-feeding brine shrimp. Suspended sediments can adsorb or desorb nutrients such as nitrogen or phosphorus, organic compounds, and toxic substances. Much more information is required about the sediments of Mono Lake and the quantity expected to be transported into the pelagic region before these possibilities can be evaluated.

TABLE 6.3 Predicted Effects of Lake Elevation and Salinity on Aquatic Plants

Lake Elevation (ft)	Salinity (g/l)	Effects on Aquatic Plants
6430-6380 (current level)	<50-89	Phytoplankton and phytobenthic algae flourish.
6370	102	Phytobenthic algal production reduced for some species.
6360	121	Reduction in phytoplankton productivity.
6350	148	Decrease in phytoplankton and phytobenthic productivity. Shift in phytoplankton species composition expected.
6340-6330	185-237	Large decreases in all types of algal productivity. Further changes in species composition.

Because of bioenergetic demands placed on larval growth and development, brine shrimp populations could be expected to gradually decrease in abundance if salinity increased above approximately 120 g/l. A decrease in hatching of dormant brine shrimp embryos would be expected if the salinity increased to approximately 130 g/l. This effect is attributable to a reduction in the "free water" of hydration of the embryos. Rapid decreases in the overall abundance of brine shrimp populations could be expected if salinity increased to approximately 150 g/l. This rapid decrease is predicted on the basis of postulated reductions in primary phytoplankton productivity, which translates into less food, slower growth of all larval stages, and less reproductive output of the adults and lack of cyst hatching. Similarly, a gradual decrease in primary phytobenthic productivity through losses of hard rock-mudflat surfaces and elevated salinities would alter the quantity and composition of benthic foodstuff.

Reduction of benthic foodstuff along with the bioenergetic demands of osmotic stress caused by increases in salinity suggests that salinities of 130 to 150 g/l would result

TABLE 6.4 Predicted Effects of Lake Elevation and Salinity on Aquatic Animals

Lake Elevation (ft)	Salinity (g/l)	Effects on Aquatic Animals
6430-6380 (current level)	<50-89	Brine shrimp and brine fly populations flourish.
6370	102	Brine shrimp populations unimpaired. Brine fly populations unimpaired by physiological effects of salinity. Loss of about 40% of submerged hard substrate relative to 6380 ft.
6360	121	Brine shrimp experience slight impairment of hatch. Brine fly larvae show modest decrease in growth.
6350	148	Brine shrimp: no hatch of cysts; decreased naupliar growth and juvenile metamorphosis; and reduced female fecundity. Brine fly: no growth of eggs; reduced female fecundity and reproductive potential.
6340	185	Brine shrimp: nonviability of dormant cysts; inhibition of preemergence mechanism of diapaused embryo; high mortalities in naupliar lifestage; and large reductions in adult populations. Brine fly: high mortalities in larval brine fly; reduced adult populations.
6330	237	Brine shrimp: loss of populations except for small populations located near freshwater spring inflows. Brine fly: loss of populations except for small populations located at shorelines where fresh water is present.

in severe reductions in the growth and development of brine fly larval and adult populations. The hard substrate required for larval grazing on benthic algae would be reduced dramatically if the lake level dropped below 6380 ft (Table 6.5). Although it has not been possible to quantitatively estimate the brine fly larval population at the current lake level, a reduction of 40 percent of the hard substrate area that would result if lake level fell to

TABLE 6.5   Area of Hard Substrate (Tufa and Mudstone)

| Lake Elevation (ft) | Hard Substrate (acres) | |
	to 10 ft below level	30 ft below level
6430	3,095	7,374
6420	1,964	7,198
6410	2,315	9,514
6400	2,919	15,452
6390	4,280	15,724
6380	8,253	13,055
6370[a]	3,191	5,655
6360	1,611	3,020
6350	853	1,860
6340	556	1,336
6330	451	958

[a]Elevations above 6370 are underestimates of hard substrates. Hard substrates north and northeast of Negit Island are not indicated because the area could not be accessed for the bathymetric survey.

SOURCE: From data in Pelagos Corporation (1987).

6370 ft (relative to the hard substrate area at the current level of 6380 ft) would certainly be detrimental to the brine fly populations.

Aquatic Bird Populations

The critical food resources for aquatic birds using Mono Lake are brine shrimp and brine flies. Eared grebes and California gulls feed primarily on brine shrimp, while red-necked phalaropes, and to a lesser extent Wilson's phalaropes, specialize on brine flies. Although we do not know the relationship between overall brine shrimp and brine fly abundance and their availability to birds in profitable concentrations, it is certain that at some point increasing salinities in the lake would cause a sufficient drop in brine shrimp and brine fly abundance that birds would be affected.   In Table 6.6 the possible impacts of changes in food supplies on grebes and phalaropes for various lake levels are predicted.   Except insofar as they affect prey populations, changes in the salinity of Mono Lake are expected to have little or no direct physiological effect on birds.

In addition to adequate food supplies, California gulls breeding at Mono Lake also require nesting sites safe from disturbance and predation.   For these sites, the gulls depend on islands in the lake, where they are safe from canids.   Island area decreases with increasing lake levels above 6380 ft as islands are inundated, and is roughly constant for lake levels between 6380 and 6350 ft.   Below 6350 ft there is a precipitous loss of island area (Table 6.7). What is harder to predict are possible changes in the habitat quality of the available island area.   The gulls have shown a willingness to change sites as land bridges form, and the committee assumes in this analysis that one area can be substituted for another with little effect.   Likewise, the figures of island area reflect the present conditions, and do not take into account the erosion of the small, unconsolidated islets that result from fluctuating, particularly rising, water levels.   To the extent that these islets are important nesting sites, their accelerated erosion could result in greater restriction of gull nesting habitat than is apparent from the overall figures of island area.   It should be noted that, if the lake level fell to the level that would significantly decrease island area, most gulls would likely have already deserted Mono Lake because of lack of food.

Mono Lake is of regional importance to populations of eared grebes, Wilson's phalaropes, and California gulls, and of local importance to red-necked phalaropes.   The lake is one of the major Great Basin staging areas used by eared grebes and Wilson's phalaropes during the fall migration. Birds currently using Mono Lake as a staging area, or even as a migratory stopover (red-necked phalaropes), may be able to shift to alternative sites, such as the Salton Sea or

TABLE 6.6   Predicted Response of Aquatic Birds to Changing Lake Levels

Lake Elevation (ft)	Response of Bird Populations
6370-6430	Migrant grebes remain at Mono Lake, molting and fattening, until brine shrimp populations become depleted. Migrant Wilson's phalaropes remain at Mono Lake until they complete their molt and become fat; red-necked phalaropes fatten during migratory stopover. Gulls breed in present numbers with present levels of success.
6360-6370	Migrant grebes leave Mono Lake earlier than they do at higher lake elevations. Wilson's phalaropes find it difficult to complete molt and premigrating fattening before departing; red-necked phalaropes' stopover shortened. Gulls may show some decreases in numbers or reproductive success due to food shortages.
6350-6360	Duration of grebes' stopover probably no longer permits molting or fat deposition. Migrant phalaropes cannot use Mono Lake as a major stopover site. Breeding numbers of gulls drastically reduced and reproductive success lowered due to lack of food.
6330-6350	Grebes and phalaropes do not use Mono Lake as a stopover site. Few gulls attempting to breed. Those present largely dependent on food resources distant from Mono Lake.

San Francisco Bay, provided such appropriate habitats continue to exist. However, the committee does not know whether these sites can sustain such a major influx of new individuals and the long-term implications of such a shift. The islands of Mono Lake provide the nesting sites of a significant proportion of the North American (world) population of California gulls. Loss or severe damage to any of the populations of eared grebes, California gulls, or phalaropes at Mono Lake would not only change strikingly the scenic features and the ecology of the Mono Basin scenic area, it may also have a potential impact on Great Basin and western North American populations of these species.

TABLE 6.7 Predicted Changes in California Gull Nesting Habitat with Changes in Lake Elevation (baseline of 6380 feet)

Lake Elevation (ft)	Response of Gull Nesting Habitat
6430	1406 acres of emergent islands, no land bridges.
6420	1439 acres of emergent islands, no land bridges.
6410	1514 acres of emergent islands, no land bridges, most Paoha islets submerged.
6400	1703 acres of emergent islands, no land bridges, many Paoha islets submerged, some loss of gull breeding habitat.
6390	2016 acres of emergent islands, no land bridges, some Paoha islets submerged, some loss of gull breeding habitat.
6380 (current level)	2655 acres of emergent islands, no land bridges, most Paoha islets emergent, 40,000 to 50,000 gulls nesting.
6370	2272 acres of emergent islands, some land bridges. Negit Island joined to shore, resulting in major loss of some currently used gull breeding area.
6360	2490 acres of emergent islands. Paoha Island still an island.
6350	2616 acres of emergent islands. As above.
6340	195 acres of emergent islands. Paoha Island joined to shore. Very little nesting area left, although depending on its character it may still be sufficient to support a large gull colony.
6330	163 acres of emergent islands. As above.

SOURCE: Island area measurements are from Pelagos Corporation (1987).

Shoreline Environment

The shoreline around Mono Lake would be inundated or exposed if lake level were to rise above or fall below the current level. The area of shoreline that would be affected at different levels is shown in Table 6.8. Consequences of these changes for the abiotic components of the shoreline environment--nearshore groundwater, tufa formations, and air quality--and the biotic components of the shoreline environment--the snowy plover habitat and the shoreline vegetation--are discussed below.

Nearshore Groundwater

If the lake level were to rise or fall, locations of springs and marshes at the lake margin would generally rise or fall in elevation accordingly. Marsh geometries and spring locations would depend on locations of tufa, joints, and high permeability zones, as well as on surface topography.

Large shoreline springs are located in the County Park area where alluvial fan deposits are recharged by Mill, Wilson, and Dechambeau creeks. Here and along the west side of the lake to the Old Marina, marshlands would migrate to lower elevations if lake level fell. The areal extent of the seepage zones may increase if the lake elevation were to fall below approximately 6380 ft because of the decreasing slope of the exposed lake bed. Similarly, at Navy Beach, where spring waters appear to be a mixture of shallow groundwater and fault-controlled flows, spring locations would migrate if lake levels changed.

With the exception of Simon's Spring, zones of seepage on the north and east sides of the lake are located away from the lakeshore, apparently occurring at the outcrop of shallow, water-bearing sediments. Unless the lake level were to rise to the elevation of these zones, seepage would not be affected by changing lake levels. But at Simon's Spring, groundwater flow is influenced by the presence of a fault that extends across the area into the lake. A comparison of past and recent aerial photographs indicates that the surface area of the marshlands has increased as lake levels have dropped and exposed more of the fault zone. If the lake level changes in the future, the shape of the marshlands would be altered. The marshlands would migrate with the shoreline and might continue to enlarge if lake level fell and might become smaller if lake level rose.

TABLE 6.8   Areas of Lake Bed Exposed at Different Elevations with Incremental Changes Between Lake Elevations (baseline at 6430 ft)

Lake Elevation (ft)	Exposed Lake Bed			Incremental Change		
	Square miles	Acres	Hectares	Square miles	Acres	Hectares
6430	0	0	0			
				4.9	3,128	1,266
6420	4.9	3,128	1,266			
				3.0	1,968	797
6410	7.9	5,096	2,063			
				3.7	2,320	939
6400	11.6	7,416	3,002			
				4.5	2,929	1,187
6390	16.1	10,346	4,189			
				6.8	4,302	1,741
6380	22.9	14,647	5,930			
				13.0	8,322	3,369
6370	35.9	22,969	9,299			
				5.5	3,526	1,428
6360	41.4	26,495	10,727			
				4.0	2,565	1,038
6350	45.4	29,060	11,765			
				4.5	2,910	1,178
6340	49.9	31,970	12,943			
				4.3	2,747	1,112
6330	54.2	34,717	14,055			

SOURCE: Data from Pelagos Corporation (1987).

Most of the shoreline north of Simon's Spring to the lakeshore southeast of Bridgeport Creek is bordered by salt flats, which have become progressively wider as lake levels have lowered. This region is characterized by low ground surface gradients and a very shallow water table. The shallow groundwater has high total dissolved solids near the lake, indicating relatively little flow toward the lake. These salt flats would probably continue to become wider if lake level were to drop and narrower if lake level were to rise.

Tufa Formations

If the lake level were to rise above the current level, tufa groves would be at least partially submerged. Many of the taller lithoid towers, however, would be visible even though their bases would be underwater. As discussed in chapter 5 and summarized in Table 6.9, elevations of tufa groves range from approximately 6368 to 6430 ft. If the lake rose to approximately 6404 ft, visitors would be able to view but would not have access to the South Tufa and Lee Vining Tufa areas. This situation would occur at lake levels of 6415 ft at the County Park and Bridgeport Creek tufa groves and at approximately 6425 ft at the Old Marina. Tufa towers occur above 6430 ft at Simon's Spring and Warm Springs and would consequently not be affected by lake elevations considered in this report.

Wave action against the base of the lithoid tufa formations could damage some of the towers, though information is not available to estimate the extent of the damage. Wide fluctuations in lake level would be more likely to cause damage to a larger number of tufa towers than would a more constant lake level.

The sand tufa, more delicate than the lithoid towers, are more likely to be significantly damaged if inundated. The sand tufa occur at elevations ranging from approximately 6390 to 6432 ft. Consequently, if the lake level were to rise above 6390 ft, some of the sand tufa would be destroyed. All of the formations would be destroyed if the lake level were to reach 6432 ft.

If the lake level fell below its current level, tufa formations that are currently submerged would be exposed. The exposure would increase the areas of tufa groves available to visitors, and could also increase the number of formations subject to vandalism. Tufa groves at South Tufa, Lee Vining Tufa, and County Park areas would be completely exposed if lake level fell to approximately 6368, 6368, and 6370 ft, respectively.

TABLE 6.9 Predicted Effects of Lake Elevation on Tufa
Formations

Lake Elevation (ft)	Effects on Tufa Formations
6430	Old Marina tufa are submerged. Tufa towers still visible. All sand tufa inundated.
6420	County Park and Bridgeport Creek tufa areas submerged. Tufa towers still visible.
6410	South Tufa and Lee Vining tufa areas submerged. Tufa towers still visible.
6400	Some additional sand tufa inundated.
6390	Some sand tufa inundated.
6380	Current level.
6370	Currently submerged tufa formations exposed at South Tufa, Lee Vining, and County Park.
6360–6330	Some currently submerged tufa formations would be exposed with declining lake levels.

NOTE: Locations of tufa groves are shown in Figures 5.8 and 5.9.

Air Quality

Falling lake level would expose more of the former lake
bed and thus presumably provide more particulates and thus
more aerosols during windstorms.   Similarly, should the
lake continue to rise, less of the former lake bed would be
exposed and, presumably, dust storms would be less intense.
A major factor in dust storms is the frequency and duration of strong wind events. These vary from one year to
another depending on storm types, frequency of frontal
passages, and the direction and strength of polar and subtropical jet streams. Simply put, some years are much
windier than others. Consequently, the concentration of
dust depends not only on the area of former lake bed exposed but also on the hydrometeorological conditions. For
example, strong wind episodes that occur when the basin is
snow-covered will raise few aerosols.

The character and moisture content of the lake bed material also affect the severity of dust storms. Saint-Amand et al. (1986) note that the physical and chemical morphology of alkali crystals creates fine dusts most susceptible to airborne transport. The potential availability of aerosols depends on future mitigation efforts, including the use of sand fences, stabilization by vegetation, and other innovative projects.

Snowy Plover Nesting Sites

As summarized in Table 6.10, the alkali flats along the eastern shore of the lake that provide the nesting for the snowy plover would be gradually inundated if lake levels rose. At levels above approximately 6420 ft the population of snowy plovers would be reduced to a few pairs nesting in elevated dunes. For lake levels between 6380 and 6420 ft, the areal extent of the alkali flats would be incrementally reduced, but information is not available to predict precisely how these changes would incrementally affect the snowy plover populations.

Though the population has benefited from decreases in lake level and exposure of the alkali flats, continued expansion of the flats is not likely to expand the snowy plover habitat because distances from the nest to the shoreline feeding areas would be too great. More importantly, access to feeding areas may be restricted with reduction in the length of the lake's shoreline with decreases in lake level. The nesting population would probably approach its maximum level at a lake elevation of approximately 6360 ft.

Shoreline Vegetation

The shoreline vegetation mosaic is directly related to the amounts of freshwater flow and the salinity of soil pore water. Areas that exemplify the influence of freshwater availability can be found at Simon's Spring, Warm Springs, along the northwestern shore, and near Gull Bath. The influence of pore water salinity is most obvious along

TABLE 6.10 Predicted Response of Snowy Plovers to Changes in Lake Elevation

Lake Elevation (ft)	Response of Snowy Plovers
6430-6410	Nesting population reduced from current level to a few pair located in elevated dunes.
6410-6380	Nesting population decreased from current level in proportion to area of exposed playa.
6380	Nesting population of approximately 350 pair.
6370-6360	Nesting population increased in proportion to area of exposed lake bed.
6360	Nesting population probably approaches maximum level.
6360-6330	No further increases in nesting population, possible decreases.

the eastern and northeastern shores of the lake, where the wind-swept salts and sands accumulate on the exposed lake bed.

Vegetation immediately associated with the springs and seeps is characteristically marshy. If the shoreline is steep, as on the western shore, the marsh vegetation may grow adjacent to the lake. However, when the lake level was lower in the late 1970s (6378 ft and lower), there was barren shoreline below the marshes in these areas. It is not clear whether these areas were barren because plants did not have sufficient time to colonize the exposed lake bed or because the salinity of the exposed soils was too high for plant establishment.

Spring or seepage areas near the lakeshore at the eastern end of the lake form localized marshy areas. In some cases the marsh areas grade into barren, alkaline lake bed through a narrow band of saltgrass (*Distichlis spicata*) and associated herbaceous vegetation (e.g., Warm Springs). In other areas where freshwater flow is sufficient to maintain vegetation over larger areas, the marsh thins toward the lake or grades into saltgrass, which continues on to the shoreline (e.g., Simon's Spring).

Two areas with vegetative cover atypical for the shore-line at Mono Lake are the alluvial fans where Lee Vining and Rush creeks enter the lake. The alluvial fans have formed terraces that drop abruptly into the lake. Because of their height above the lake and the supply of fresh groundwater from the Sierra, they support a shrub com-munity dominated by rabbitbrush or grassy vegetation. Sparse stands of willows and other riparian species are found in the river channels.

The changes in shoreline vegetation that could be ex-pected with changes in lake level are summarized in Table 6.11. Generally, the changes in vegetation would be con-trolled by the changes in fresh water available from spring or seeps, as described in the above sections on nearshore groundwater.

Upland Environment

If the lake level were to be maintained at any given level, specific amounts of water would have to flow into the lake each year. Based on the modeling results described in chapter 2, necessary inflows required to main-tain given lake levels are reported as average flows at the diversion points (Table 6.12). The consequences of chang-ing streamflows for the riparian vegetation and riparian habitats are discussed below.

Riparian Vegetation

Riparian vegetation is dependent on more soil moisture than is supplied by normal percolation of rain or snow fall. For this reason, riparian species typically occur along stream courses, on the lake edge, and where groundwater reaches the surface at springs and seeps. Reduction or elimination of the supplemental water source from surface runoff or from lakes and springs may stress riparian spe-cies so severely that they will be eliminated from the system.

The riparian systems of streams that enter Mono Lake are discussed in chapter 5. Four of these streams have

TABLE 6.11 Predicted Shoreline Responses to Changes in Lake Elevation (baseline of 6380 ft)

Lake Elevation (ft)	Shoreline Responses
6430	Continued inundation of grassland flats and barren shore and possible local upward migration of marshes.
6400	Inundation of barren shores on northeast, east, and southeast, phreatophytic saltgrass flats and spring areas
6390	Inundation in nonvegetated areas on east of lake, reducing barren playas. Inundation of many phreatophytic grass areas. Inundation of marshes on west end steep shoreline. Possible upward migration of marshes on western shore.
6380	Present status--baseline.
6370	Eastern shoreline recedes by 4000-6000 surface ft and up to 8000 ft near Negit Island. Annual plants will invade black sands along north shore. Most exposed lake bed will remain barren. Warm Springs will continue to maintain a marsh well above shoreline. Simon's Spring as a source of fresh water may cause encroachment of saltgrass and other phreatophytic herbs along the fault that controls the spring. Tufa areas on south central and western shores may become freshwater spring areas and be invaded by marsh vegetation. Marshes on northwestern end of lake may gradually invade shore where fresh water percolates to surface. Upland vegetation may migrate downward. Steep shorelines near Lee Vining and Rush creek deltas and around islands will erode and may prevent vegetation establishment.
6360	Most exposed lake bed will remain barren, especially northeast, east, and south shores. Simon's Spring seepage may continue to allow localized encroachment of saltgrass onto exposed lake bed along fault. West and northwest shores may still have sufficient freshwater seepage to permit marsh establishment on exposed lake bed. Tufa areas may have sufficient springs to support herbaceous vegetation. Rush, Lee Vining, and other creeks will be deeply incised with upstream downcutting eliminating established vegetation and denuding banks; stream deltas will be expanded.
6350	Newly exposed lake bed probably will support no vegetation; it may be too saline. Exceptions to barren lake bed may be along Simon's Spring fault and tufa areas. Surface erosion from windblown particles will become more critical to vegetation establishment, especially along eastern end of lake.
6340-6330	Newly exposed shoreline will be barren.

TABLE 6.12 Approximation of the Instream Flow at Diversion Points Required to Maintain a Selected Level of Mono Lake

Lake Elevation (ft)	Instream Flow at Diversion Points				
	LADWP model		Vorster model		
	Acre-Feet per Year	Cubic Feet per Second	Acre-Feet per Year	Cubic Feet per Second	
6430	*	*	*	*	
6420	*	*	*	*	
6410	*	*	*	*	
6400	102,977	142	*	*	
6390	80,199	111	91,824	127	
6380	58,464	81	79,976	110	
6370	38,011	53	67,602	93	
6360	21,300	29	54,847	76	
6350	10,318	14	42,548	59	
6340	*	*	30,528	42	
6330	*	*	19,172	26	

*Out of range of the formatted models' calculated results.

had a major part of their instream flows diverted for over 40 years. Riparian vegetation below diversions on those streams has suffered and little remains.

Maintenance of various lake levels above the stabilization level will require input of various amounts of surface water into these stream systems as shown in Table 6.12. The response of riparian species to any given input is dependent on water requirements for germination, successful establishment of seedlings, and survival of plants during the maturing process.

Streamflow apparently does not have to be perennial to maintain a riparian community; many intermittent streams in the semiarid Southwest support riparian vegetation of varying densities. Therefore, some streamflow should maintain some riparian vegetation, although limited flow may never create the environment that will stimulate recruitment or regeneration of some riparian species.

With high flows in Rush Creek in 1983 and periodically since, and with minimum flows of 19 cfs since 1985, the surviving riparian vegetation has been rejuvenated. Isolated willow and *Populus* plants have established on the gravel bars along the stream. This surface flow regime in Rush Creek has apparently been sufficient to stimulate reestablishment of woody riparian vegetation.

Below the water diversion points, Rush and Lee Vining creeks cross alluvial deposits. That material is so porous that it does not hold groundwater near enough to the surface and is inadequate to maintain riparian plants if the upstream input is low. Of the 19 cfs released into Rush Creek at Grant Lake, some but not all reaches Mono Lake. Some of the flow recharges below-surface aquifers that could be used by riparian vegetation. Evidence that the released flow is adequate to recharge the channel alluvium is seen in the fact the water reaches the lake via the stream channel. In 1986, a continual flow of 10 cfs was released into the lower reaches of Lee Vining Creek. That amount is apparently adequate to recharge the channel alluvium and maintain enough flow to reach the lake. Again, the flow reaching the lake was undoubtedly less than 10 cfs, but, as at Rush Creek, it was not measured.

The streamflows given in Table 6.12 would be measured at the point of release and do not represent the amount of water entering the lake. With little or no riparian vegetation along the lower reaches of Rush and Lee Vining creeks, the relationship between a flow at the diversion points and the flow entering the lake will remain constant; however, once riparian vegetation becomes reestablished, the flow into the lake will decrease because of increased loss from evapotranspiration.

If the 7-mi stretch of Rush Creek between Grant Lake and Mono Lake had a riparian strand averaging 750 ft wide (approximate average width measured from pre-1941 from aerial photos) and was composed of two-thirds *Populus* species and one-third willow, the annual evapotranspiration loss would be about 3500 acre-ft based on information in chapter 5. That loss is equivalent to a reduction in streamflow of about 4.8 cfs, an amount that may maintain a riparian strand of this size if none of the streamflow were lost into the porous substrate.

The minimum flows of 19 and 10 cfs currently maintained in Rush and Lee Vining creeks should be adequate to maintain riparian strands equivalent to those existing in 1941. According to the model results discussed in chapter 2, these flows (29 cfs) would be the average flow to maintain the lake level at 6360 ft as predicted by the LADWP (1987) model. Vorster's model (Vorster, 1985) predicts that the flows would be the average to maintain the lake level at approximately 6330 ft (Table 6.12). In light of the uncertainties in the model predictions and the committee's conservative approach to predicting effects of changes in streamflow on the riparian strand, the committee concludes that flows necessary to maintain lake levels above 6360 ft, regardless of which model is used, should maintain riparian strands on Rush and Lee Vining creeks. From another perspective, if minimal releases of 19 cfs for Rush Creek and 10 cfs for Lee Vining Creek are maintained, the composite (29 cfs) is predicted to maintain lake levels of 6360 or 6330 ft, depending on which model is used (Table 6.12). Intermittent dry periods and changes in flow probably would have little negative effect, especially in the fall and winter. However, maintenance of riparian vegetation would also require periodic heavy releases in the spring to enhance the recruitment potential of riparian species.

Releases of less than 29 cfs may not be sufficient to maintain a riparian community. If released solely in Rush Creek, the flow may be sufficient to maintain a riparian strand. If the flow were to be divided between Rush and Lee Vining creeks, however, it would probably be inadequate to maintain a healthy riparian strand.

The role of a riparian strand becomes important in calculating the flow required to maintain any given lake level. If the fully reestablished riparian strand along Rush Creek were to consume 3500 acre-ft/yr of water, the lake levels that could be maintained by a given release would be reduced by about 2 ft. Establishment of a riparian strand along Lee Vining Creek would lead to yet further reduction in lake level.

Fish

Because no fish live in Mono Lake itself, the water level in the lake has no direct effect on any fish populations in the basin. However, the water level in Mono Lake is affected by the amount of water flowing in the streams that drain the basin (Table 6.12). Although it is difficult to predict the relationship between flow at the diversion points and the suitability of the stream for fish if the variation in flow is not known, it is certain that sustained moderate releases spread throughout the year would be more beneficial to fish populations and the animals that they rely on for food than a single major release in spring followed by a complete lack of flow during the remainder of the year. Similar considerations apply to mammalian wildlife that uses the riparian systems.

At present, minimal flows are being maintained of 19 cfs in lower Rush Creek and 10 cfs in lower Lee Vining Creek. These flows are adequate to support reproducing populations of brown trout in the two streams, and some rainbow trout may be reproducing in lower Lee Vining Creek. It is probable that increasing the flows (up to a point) would increase the sizes of the trout populations and the number of trout species that reproduce in Lee Vining and Rush creeks.

SUMMARY AND CONCLUSIONS

The resources of the Mono Basin ecosystem--the aquatic biological community, bird populations, and shoreline and upland environments--are affected by changes in lake level in different ways. Some would be adversely affected if lake level rose above the current level (6380 ft), and others would be adversely affected by lower lake levels. In using the conclusions of this report to determine how the ecosystem should be managed, decisions about the relative importance of different resources will have to be made.

It is important to keep in mind that the responses of the various resources to changes in lake level would occur gradually over a range of levels. A precise lake level at which an impact would occur cannot generally be pinpointed. Rather, the impacts would increase in intensity with changes in lake level to a point where the impact becomes severe.

The effects of changes in lake level are summarized in Table 6.13, and the range of levels over which the impacts occur are shown in Figure 6.3. The major ecological concerns with changes in lake level involve the ecological effects of salinity and habitat availability.

If the lake level were to drop, salinity would increase and reduce the water available for metabolism ("free water") for the dominant aquatic organisms: algae, brine shrimp, and brine flies. This reduction would increase the physiological costs of reproduction and growth to the organisms. At salinities around 120 g/l, populations of these organisms would begin to show negative responses, and at 150 g/l acute ecological effects--drastic population reductions--are predicted. These salinities are associated with lake levels of approximately 6360 and 6350 ft, respectively.

The chemical stratification of the lake, known as meromixis, that currently occurs and can be expected to occur in the future, particularly if the lake level were to fall, results in an upper layer of water that is less saline than the bottom layer. Therefore, the adverse effects of increases in salinity would be alleviated to some degree.

If salinity increased, precipitation of salts would occur and would increase the probability and persistence of meromixis.   Although the precise effects of precipitation on salinity cannot be quantified with the current state of understanding about the geochemistry of Mono Lake, the salts would begin to precipitate at salinities above approximately 125 g/l.   The precipitated salts would reduce the salinity but would generate a dense layer toward the lake bottom that would resist mixing with the overlying water column. This layer would act as a sink for nutrients that would no longer support biological productivity through vertical mixing.

The committee concludes that, in balance, the relationship between lake level and salinity predicted assuming a constant quantity of evenly distributed salt in the lake results in an upper limit on the estimate of salinity for a given lake level. Consequently, predictions about the effects of falling lake levels on the organisms assume the worst case.   This conservative prediction is appropriate when considering the uncertainties of the data and the severity of the consequences of increases in salinity for the Mono Basin ecosystem.

If the lake level dropped, there would be a loss of littoral habitat that is essential to the phytobenthic productivity and the brine flies.   If lake level fell 10 ft from its current level, the amount of hard substrate available as littoral habitat for brine flies would be reduced by 40 percent.

The nesting and migratory bird populations at Mono Lake are affected by changes in lake level through changes in food chain productivity and availability of nesting habitat.   The decreases in availability of brine shrimp for food would begin to have adverse effects on those bird species relying on brine shrimp--eared grebes and California gulls--at salinities of 120 g/l (6360 ft).   The impacts would be acute at 150 g/l (6350 ft).   For those birds relying on brine flies--the phalaropes--impacts would begin to become apparent at 6370 ft and would be acute at 6360 ft. The nesting habitat for the California gulls is adversely affected by emerging land bridges and positively affected by new island emergence as water level decreases.   A lake level of 6340 ft represents the breakpoint where further reductions are associated with major losses in nesting habitat.

TABLE 6.13   Predicted Major Effects of Changing Lake
Elevation on Resources of Mono Basin   (Effects are relative to current elevation of 6380 ft)

Resource	Major Effect of Changing Lake Elevation
Aquatic biology	
Algae	Large reductions in productivity, caused by reduced nutrient supply associated with meromixis, increasingly likely at salinities above 125 g/l (corresponding to lake elevation of approximately 6358 ft).
	Large reductions in productivity and species shifts related to salinity increases above 150 g/l (corresponding to lake elevation of 6350 ft) for benthic algae and above 185 g/l for phytoplankton.
Brine shrimp	Physiological effects of increase in salinity slight at 120 g/l (corresponding to lake elevation of 6360 ft) and severe at 150 g/l (corresponding to lake elevation of 6350 ft).
Brine fly	Hard substrate required for habitat substantially reduced if lake elevation falls below 6370 ft. Physiological effects of increase in salinity severe at 150 g/l.
Bird populations	
Eared grebe	Affected by decreases in availability of brine shrimp at 6350 to 6360 ft.
Phalaropes	Affected by decreases in availability of brine fly at 6360 to 6370 ft.
California gull	Affected by decreases in availability of brine shrimp at 6350 to 6360 ft. Island area for nesting severely affected at elevations below 6350 ft.
Snowy plover	Gradual reduction in shoreline nesting area if lake elevation rises. Nesting population area approaches maximum size at 6360 ft.
Shoreline environment	
Vegetation	Increases in lake elevation gradually inundate saltgrass. Decreases in lake elevation would extend exposed lake beds. Vegetation would be established only in area with springs.
Air quality	Increases in lake elevation would inundate alkali flats and reduce dust problem. Decreases in lake elevation would expose alkali flats and exacerbate problem.
Tufa	Increases in lake elevation would gradually inundate tufa groves to 6430 ft. Sand tufa gradually destroyed with increases in elevation from 6410 to 6430 ft. Decreases in lake elevation would expose

TABLE 6.13 (continued)

Resource	Major Effect of Changing Lake Elevation
Upland environment	more tufa for tourism. Exposed tufa may be susceptible to vandalism.
Riparian vegetation	Currently maintained flows of 19 cfs in Rush Creek and 10 cfs in Lee Vining (approximate average flow at diversion points to maintain lake elevation at 6360 ft or 6330 ft depending on the model used) adequate to maintain healthy riparian strand. Periodic flooding needed for riparian species recruitment.
Riparian habitat	Currently maintained flows adequate to maintain healthy fish populations.

The exposure of the lake bed with lower lake levels would generate an increase in the areal extent of the shoreline playas and a lower water table under much of the existing shoreline vegetation. Exposure of playas would increase the area available for the snowy plover to nest. The maximum nesting area would be approached at lake levels around 6360 ft. Inundation of salt flats if lake level rose would reduce the area available to the snowy plover.

If lake levels dropped, existing tufa formations would be exposed to more erosion and human damage, though submerged tufa formations would be exposed for visual enjoyment. On the other hand, increases in lake level would inundate currently exposed tufa formations, though the formations would still be visible. A widely fluctuating lake level would lead to more damage from wave action than would a lake level with less fluctuation. The sand tufa would be gradually destroyed if the lake level were to increase from 6410 to 6430 ft. The increase in exposed lake bed would exacerbate existing downwind air quality problems.

Alterations in the flow of the diverted streams in the basin would be expected with changes in lake level, although it is not possible to predict the precise flow regime that would be associated with a particular lake level because the distribution of flows between the diverted streams depends on how the water is managed. Changes in

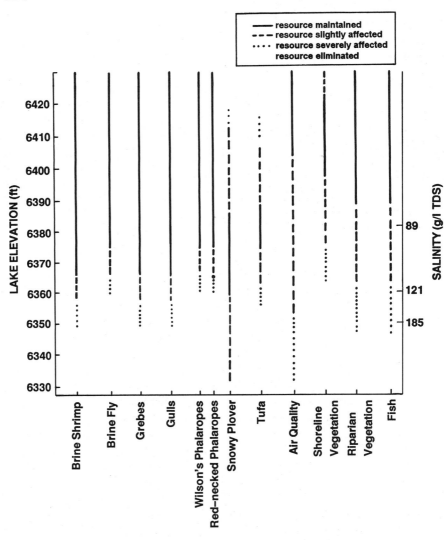

FIGURE 6.3 Ranges of lake levels affecting resources of the Mono Basin, with three salinities added for reference.

streamflow have consequences for the riparian vegetation and habitat. The flows currently maintained in Rush and Lee Vining creeks, 19 and 10 cfs, respectively, appear adequate to maintain a healthy riparian community. Depending on which of the two hydrologic models of the basin is used, these flows are estimated to be approximately the average flows at the diversion points necessary to maintain the lake level at 6360 or 6330 ft.

Determination of lake levels that would preserve the integrity of the ecosystem must take into account the need for an increment of lake level higher (or lower) than the basic minimum (or maximum) that causes catastrophic disruption of the ecosystem. This buffer is required for two reasons. First, the predictions in this report are based on data that in some cases are incomplete or uncertain, resulting in the need for a margin of safety against disruption of the ecosystem. Second, natural fluctuations in climate can cause fluctuations around a particular lake level that could not be controlled.

The predictions included in this report are the most accurate that the committee could achieve based on the available information. These predictions need to be verified by continued and expanded research and monitoring, as pointed out throughout the report.

REFERENCES

Los Angeles Department of Water and Power. 1986. Report on Mono Lake Salinity. Los Angeles, Calif.

Los Angeles Department of Water and Power. 1987. Mono Basin Geology and Hydrology. Los Angeles, Calif.

Melack, J. M. 1985. Interactions of detrital particulates and plankton. Hydrobiologia 125:209-220.

Pelagos Corporation. 1987. A Bathymetric and Geologic Survey at Mono Lake, California. Report prepared for Los Angeles Department of Water and Power. San Diego, Calif.

Saint-Amand, P., L. A. Mathews, C. Gaines, and R. Reinking. 1986. Dust Storms from Owens and Mono Valleys, California. NWC TP6731. China Lake, Calif.: Naval Weapons Center. 79 pp.

Vorster, P. 1985. A Water Balance Forecast Model for Mono Lake, California. Master's thesis, California State University, Hayward. Earth Resources Monograph No. 10. San Francisco, Calif.: U.S. Forest Service, Region 5.

Appendixes

Appendix A

Birds of the Mono Basin
and Their Ecological Characteristics

LEGEND: Habitat (1 = lake, 2 = exposed lake bed, 3 = marsh, 4 = riparian zones, 5 = sagebrush-steppe, 6 = conifer forests, 7 = alpine); Season (1 = spring (AMJ), 2 = summer (JAS), 3 = fall (OND), 4 = winter (JFM)); Status (1 = summer resident, 2 = migrant or visitor, 3 = winter resident); Peak Populations (1 = 10s, 2 = 100s, 3 = 1000s, 4 = 10,000s, 5 = 100,000s); Response to Disturbances (A = lowering lake level, B = decreasing streamflows, C = controlling fire, D = increasing grazing), 0 = no response likely, + = increased population, - = decreased population.

Species	Habitat	Season	Status	Peak Population Size	Response to Disturbances			
					A	B	C	D
Red-throated Loon	1	2,4	2	1	0	0	0	0
Arctic Loon	1	4	2	1	0	0	0	0
Common Loon	1	1-4	1,2	1	0	0	0	0
Pied-billed Grebe	1	1-3	1,2	1-2	0	0	0	0
Horned Grebe	1	1-3	1,2	1	0	0	0	0
Eared Grebe	1	1-4	1,2,3	2-5	0	0	0	0
Western Grebe	1	1-4	2	1	0	0	0	0
White Pelican	1	1-4	2	1	0	0	0	0
Double-crested Cormorant	1	1-3	2	1	0	0	0	0
Pelagic Cormorant	1	3	2	1	0	0	0	0
American Bittern	3	1-3	1,2	1	-	-	0	0
Least Bittern	3	2	1	1	-	-	0	0
Great Blue Heron	1-5	1-3	1,2	1	-	-	-	-

215

Species	Habitat	Season	Status	Peak Population Size	Response to Disturbances			
					A	B	C	D
Great Egret	1,3	1-3	2	1	-	-	0	0
Snowy Egret	1,3	1,2	2	1	-	-	0	0
Cattle Egret	5	1-4	2	1	0	0	-	+
Green-backed Heron	1,3	1-3	2	1	-	-	0	0
Black-crowned Night-Heron	3	1,2	2	1	-	-	0	0
White-faced Ibis	1	1,2	2	1	-	-	0	0
Tundra Swan	1	3-4	2	1	0	0	0	0
Trumpeter Swan	1	3	2	1	0	0	0	0
Snow Goose	1	3-4	2	1	0	0	0	0
Ross' Goose	1	3	2	1	0	0	0	0
Greater White-fronted Goose	1	1,3	2	1	0	0	0	0
Canada Goose	1-3,5	1-4	2	1-2	-	-	-	-
Brant	1	1	2	1	0	0	0	0
Green-winged Teal	1,3	1-4	1-3	3	-	-	0	0
Mallard	1,3	1-4	1-3	3	-	-	0	0
Northern Pintail	1,3	1-4	1-3	3	-	-	0	0
Blue-winged Teal	1,3	1,2	2	1	-	-	0	0
Cinnamon Teal	1,3	1-4	1,2	3	-	-	0	0
Northern Shoveler	1,3	1-4	1-3	3	-	-	0	0
Gadwall	1,3	1-4	1-3	3	-	-	0	0
American Wigeon	1	1-4	1-3	2	0	0	0	0
Canvasback	1	3	2	1	0	0	0	0
Redhead	1	1-4	2	1	0	0	0	0
Ring-necked Duck	1	1-3	2	2	0	0	0	0
Greater Scaup	1	1	2	1	0	0	C	0
Lesser Scaup	1	1,3	2	1	0	0	0	0
Black Scoter	1	3	2	1	0	0	0	0

Species	Habitat	Season	Status	Peak Population Size	Response to Disturbances			
					A	B	C	D
Surf Scoter	1	3	2	1	0	0	0	0
White-winged Scoter	1	3	2	1	0	0	0	0
Common Goldeneye	1	3	2	1	0	0	0	0
Bufflehead	1	1,3	2	2	0	0	0	0
Hooded Merganser	1	3	2	1	0	0	0	0
Common Merganser	1	1,3	2	2	0	0	0	0
Red-breasted Merganser	1	1,3	2	1	0	0	0	0
Ruddy Duck	1	1-4	1-3	3	0	0	0	0
Turkey Vulture	1-6	1-4	2	1	0	0	-	+
Osprey	1	1,3	2	1	0	-	0	0
Mississippi Kite	4	1	2	1	0	-	0	0
Bald Eagle	1,4	1,3,4	2	1	0	-	0	0
Northern Harrier	3,5	1-4	2	1	0	-	-	-
Sharp-shinned Hawk	4,5,6	1-4	1-3	1	0	-	+	-
Cooper's Hawk	4,5,6	1-4	1-3	1	0	-	+	-
Northern Goshawk	4,6	1-4	1-3	1	0	-	+	-
Red-shouldered Hawk	4	1-3	2	1	0	-	0	0
Swainson's Hawk	4,5	1,2	2	1	0	0	-	-
Red-tailed Hawk	4,5,6	1-4	1-3	2	0	-	-	-
Rough-legged Hawk	3,5	3,4	2	1	0	0	-	-
Golden Eagle	4,5,6	1-4	1-3	1	0	0	-	-
American Kestrel	3,4,5,6	1-4	1-3	2	0	-	-	-
Merlin	3,4,5,6	1,3,4	2	1	0	0	-	-
Peregrine Falcon	2-5	1,3	2	1	0	0	-	-
Prairie Falcon	2-5	1-4	1-3	1	0	-	-	-
Chuckar	5	1-4	1,3	2	0	0	-	-
Blue Grouse	6	1-4	1,3	2	0	0	+	-
White-tailed Ptarmigan	5,7	1-4	1,3	2	0	0	-	-

Species	Habitat	Season	Status	Peak Population Size	Response to Disturbances			
					A	B	C	D
Sage Grouse	5	1-4	1,3	2	0	0	-	-
California Quail	4,5	1-4	1,3	1	0	-	-	-
Mountain Quail	4,6	1-4	1,3	2	0	-	-	-
Virginia Rail	3	1-4	2	1	-	-	0	0
Sora	1-2	1,2	2		-	-	0	0
Common Moorhen	3	1	2	1	-	-	0	0
American Coot	1-3	1-4	1-3	3	-	-	0	0
Black-bellied Plover	1,2	1-3	2	1	+	-	0	0
Lesser Golden Plover	1,2	1-3	2	1	+	-	0	0
Snowy Plover	1,2	1,2	1,2	2	+	-	0	0
Semipalmated Plover	1,2	1-3	2	2	+	-	0	0
Killdeer	1,2,3,5	1-4	1-3	2	+	-	-	+
Mountain Plover	1,2	2	2	1	+	-	0	0
Black-necked Stilt	1,2	1,2	2	1	+	-	0	0
American Avocet	1,2	1,2	2	3	+	-	0	0
Greater Yellowlegs	1,2	1,2	2	2	+	-	0	0
Lesser Yellowlegs	1,2	1,2	2	1	+	-	0	0
Solitary Sandpiper	1,2,3	1,2	2	1	-	-	0	0
Willet	1,2	1,2	2	1	+	-	0	0
Wandering Tattler	1,2	1,2	2	1	+	-	0	0
Spotted Sandpiper	1,2,3,4	1,2	1,2	2	-	-	0	0
Whimbrel	1,2	1,2	2	1	+	-	0	0
Long-billed Curlew	1,2	1,2	2	1	+	-	0	0
Marbled Godwit	1,2	1,2	2	1	+	-	0	0
Ruddy Turnstone	1,2	1,2	2	1	+	-	0	0
Red Knot	1,2	1,2	2	1	+	-	0	0
Sanderling	1,2	1,2	2	1	+	-	0	0
Semipalmated Sandpiper	1,2	1,2	2	1	+	-	0	0

Species	Habitat	Season	Status	Peak Population Size	Response to Disturbances			
					A	B	C	D
Least Sandpiper	1,2	1-4	2	3	+	-	0	0
White-rumped Sandpiper	1,2	1	2	1	+	-	0	0
Baird's Sandpiper	1,2	1,2	2	1	+	-	0	0
Pectoral Sandpiper	1,2,3	2,3	2	1	+	-	0	0
Dunlin	1,2,3	1,3	2	2	+	-	0	0
Short-billed Dowitcher	1,2,3	1-3	2	2	+	-	0	0
Long-billed Dowitcher	1,2,3	1-3	2	2	+	-	0	0
Common Snipe	1-5	1-4	1-3	3	+	-	-	-
Wilson's Phalarope	1	1,2	2	4	-	0	0	0
Red-necked Phalarope	1	1,2	2	4	-	0	0	0
Red Phalarope	1	2,3	2	1	-	0	0	0
Pomarine Jaeger	1	2	2	1	0	0	0	0
Parasitic Jaeger	1	1,2	2	1	0	0	0	0
Long-tailed Jaeger	1	2	2	1	0	0	0	0
Bonaparte's Gull	1	1-3	2	1	0	0	0	0
Ring-billed Gull	1	1-4	2	1	0	0	0	0
California Gull	1-5	1-4	1-3	4	-	-	-	+
Herring Gull	1	3-4	2	1	0	0	0	0
Sabine's Gull	1	3	2	1	0	0	0	0
Caspian Tern	1,4	1,2	2	1	0	0	0	0
Common Tern	1	1,2	2	1	0	0	0	0
Forster's Tern	1	1,2	2	1	0	0	0	0
Black Tern	1	1,2	2	1	0	0	0	0
Marbled Murrelet	1	2	2	1	0	0	0	0
Band-tailed Pigeon	4,5,6	1,2	2	1	0	-	-	-
White-winged Dove	4	1	2	1	0	-	0	0
Mourning Dove	4,5,6	1-4	1-3	2	0	-	-	0
Yellow-billed Cuckoo	4	1	2	1	0	-	0	0

Species	Habitat	Season	Status	Peak Population Size	Response to Disturbances			
					A	B	C	D
Common Barn-Owl	3,4,5	1-3	2	1	0	0	-	-
Flammulated Owl	5,6	1,2	1,2	1	0	0	+	0
Western Screech Owl	4	1,3	2,3	1	0	-	0	0
Great Horned Owl	4,5,6	1-3	1,2	2	0	-	+	-
Northern Pigmy Owl	4,5,6	1-4	1,3	1	0	-	+	-
Burrowing Owl	5	2,3	2	1	0	0	+	+
Long-eared Owl	4,5,6	1-4	1-3	1	0	-	+	-
Short-eared Owl	3,5	1-3	2	1	0	0	-	+
Northern Saw-whet Owl	4,6	1-3	2	1	0	-	+	0
Common Nighthawk	3,5	1,2	1,2	2	0	0	0	-
Common Poorwill	5	1,2	1,2	2	0	0	0	-
Black Swift	1-7	1,2	2	1	0	0	0	0
Chimney Swift	1-7	1	2	1	0	0	0	0
Vaux's Swift	1-7	1	2	1	0	0	0	0
White-throated Swift	1-7	1,2	1,2	1	0	0	0	0
Black-chinned Hummingbird	4,5	1,2	2	1	0	-	0	0
Anna's Hummingbird	4,5	2	2	1	0	-	0	0
Calliope Hummingbird	4,5,6	1,2	1,2	2	0	-	-	-
Broad-tailed Hummingbird	4,5	1,2	2	1	0	-	0	0
Rufous Hummingbird	4,5,6	1,2	1,2	2	0	-	-	-
Belted Kingfisher	1,3,4	1-4	1,2,3	1	0	-	0	0
Lewis' Woodpecker	4,6	1-3	1,2	2	0	-	-	-
Acorn Woodpecker	6	2	2	1	0	0	-	-
Yellow-bellied Sapsucker	4	2,3	2	1	0	-	+	0
Red-breasted Sapsucker	4,6	1-4	1-3	3	0	-	+	0
Williamson's Sapsucker	6	1-4	1-3	1	0	0	+	0
Nuttall's Woodpecker	4	2,3	2	1	0	-	+	0
Downy Woodpecker	4	1-4	1-3	2	0	-	+	0

Species	Habitat	Season	Status	Peak Population Size	Response to Disturbances			
					A	B	C	D
Hairy Woodpecker	4,6	1-4	1-3	3	0	–	+	0
White-headed Woodpecker	6	1-4	1-3	1	0	0	+	0
Black-backed Woodpecker	6	1-4	1-3	1	0	0	+	0
Northern Flicker	4,5,6	1-4	1-3	3	0	–	+	–
Olive-sided Flycatcher	4,6	1,2	1,2	2	0	–	–	0
Western Wood Pewee	4,6	1,2	1,2	3	0	–	+	0
Hammond's Flycatcher	4,6	1,2	2	1	0	–	+	0
Dusky Flycatcher	4,5,6	1,2	1,2	3	0	–	–	–
Willow Flycatcher	4	1,2	1,2	1	0	–	–	0
Gray Flycatcher	5	1,2	1,2	2	0	0	–	–
Western Flycatcher	4,6	1,2	1,2	2	0	–	–	0
Black Phoebe	5	1	2	1	0	0	0	0
Say's Phoebe	5,6	1-3	1,2	2	0	0	–	0
Vermillion Flycatcher	3	3	2	1	0	0	0	0
Ash-throated Flycatcher	4,5,6	1,2	2	1	0	0	0	0
Western Kingbird	4,5	1,2	2	1	0	0	0	0
Horned Lark	5	1-4	1-3	3	0	0	–	+
Tree Swallow	4	1,2	1,2	3	0	–	–	0
Violet-green Swallow	4,6	1,2	1,2	3	0	–	–	0
Bank Swallow	1,3	1,2	2	1	0	0	0	0
Northern Rough-winged Swallow	4	1,2	1,2	2	0	–	–	0
Cliff Swallow	1-5	1,2	1,2	3	0	–	–	0
Barn Swallow	1-5	1-3	1,2	3	0	–	–	0
Steller's Jay	4,5,6	1-4	1-3	3	0	–	+	–
Scrub Jay	5	1-4	1-3	2	0	0	–	–
Pinyon Jay	5,6	1-4	1-3	2	0	0	–	–
Clark's Nutcracker	6	1-4	1-3	3	0	0	+	–

Species	Habitat	Season	Status	Peak Population Size	Response to Disturbances			
					A	B	C	D
Black-billed Magpie	5	1-4	1-3	2	0	-	-	-
American Crow	5	1,3	2	1	0	0	0	0
Common Raven	1-6	1-4	1-3	2	0	-	-	+
Mountain Chickadee	4,6	1-4	1-3	3	0	-	-	0
Plain Titmouse	5,6	1-4	1-3	2	0	0	+	0
Bushtit	5,6	1-4	1-3	2	0	0	+	0
Red-breasted Nuthatch	6	1-4	1-3	2	0	0	+	0
White-breasted Nuthatch	6	1-4	1-3	3	0	0	+	0
Pygmy Nuthatch	6	1-4	1-3	3	0	0	+	0
Brown Creeper	4,6	1-4	1-3	3	0	-	+	0
Rock Wren	5,6	1,2	1-3	2	0	-	+	0
Canyon Wren	5,6	1-4	1-3	1	0	-	+	0
Bewick's Wren	4,5	1-4	2,3	2	0	-	+	0
House Wren	4,5	1,2	1,2	3	0	-	-	+
Winter Wren	4	1,3,4	2	1	0	0	+	0
Marsh Wren	3	1-4	1-3	3	-	-	0	0
American Dipper	1,4	1-4	1-3	2	-	-	0	0
Golden-crowned Kinglet	4,6	1-4	1-3	2	0	-	+	0
Ruby-crowned Kinglet	4,5,6	1-3	1,2	3	0	-	+	-
Blue-gray Gnatcatcher	5	1,2	1,2	2	0	0	-	-
Mountain Bluebird	5,6	1-4	1-3	2	0	0	-	-
Townsend's Solitair	4,6	1-4	1-3	2	0	-	+	0
Swainson's Thrush	4	1,2	1,2	1	0	-	+	0
Hermit Thrush	6	1-3	1,2	2	0	0	+	0
American Robin	4,5,6	1-4	1-3	3	0	-	-	-
Varied Thrush	4,6	1-3	2	1	0	0	0	0
Northern Mockingbird	5	1,2	2	1	0	0	0	0
Sage Thrasher	5	1,2	1,2	2	0	0	-	-

Species	Habitat	Season	Status	Peak Population Size	Response to Disturbances			
					A	B	C	D
Water Pipit	1,2,5	1-3	1,2	2	-	-	0	-
Bohemian Waxwing	6	3,4	2	1	0	0	0	0
Cedar Waxwing	4,5,6	1-3	2	1	0	0	0	0
Phainopepla	4,5	3	2	1	0	0	0	0
Northern Shrike	4,5,6	3,4	2	1	0	0	0	0
Loggerhead Shrike	5	1-3	1,2	1	0	0	-	-
European Starling	2,4,5	1-4	1-3	2	0	-	-	+
Solitary Vireo	4,6	1,2	2	1	0	0	0	0
Warbling Vireo	4,6	1,2	1,2	2	0	-	+	0
Tennessee Warbler	4	1,3	2	1	0	0	0	0
Orange-crowned Warbler	4,5	1-3	1,2	2	0	-	+	0
Nashville Warbler	4,5,6	1,2	2	1	0	0	0	0
Virginia's Warbler	4,6	1,2	2	1	0	0	0	0
Northern Parula	4	1	2	1	0	0	0	0
Yellow Warbler	4	1,2	1,2	3	0	-	+	-
Chestnut-sided Warbler	4	2	2	1	0	0	0	0
Yellow-rumped Warbler	4,5,6	1-3	1,2	3	0	-	+	-
Black-throated Gray Warbler	4,6	1,2	2	1	0	0	0	0
Townsend's Warbler	4,6	1,2	2	1	0	0	0	0
Hermit Warbler	4,6	1,2	2	1	0	0	0	0
Prairie Warbler	4	1,2	2	1	0	0	0	0
Palm Warbler	4	1	2	1	0	0	0	0
Blackpoll Warbler	4	2	2	1	0	0	0	0
Black-and-white Warbler	4	1,2	2	1	0	0	0	0
American Redstart	4	1,2	2	1	0	0	0	0
Ovenbird	4	1,2	2	1	0	0	0	0
MacGillivray's Warbler	4,5	1,2	1,2	2	0	-	-	-

Species	Habitat	Season	Status	Peak Population Size	Response to Disturbances			
					A	B	C	D
Common Yellowthroat	3,4	1,2	2	2	0	-	0	0
Hooded Warbler	4	1,3	2	1	0	0	0	0
Wilson's Warbler	4,6	1,2	1,2	3	0	-	+	0
Yellow-breasted Chat	4	1,2	1,2	1	0	-	+	0
Summer Tanager	4	2	2	1	0	0	0	0
Western Tanager	4,5,6	1,2	1,2	2	0	-	+	0
Rose-breasted Grosbeak	4	1,2	2	1	0	0	0	0
Black-headed Grosbeak	4,6	1,2	1,2	2	0	-	+	0
Lazuli Bunting	4,5	1,2	1,2	1	0	-	-	-
Indigo Bunting	4	1,2	2	1	0	0	0	0
Green-tailed Towhee	5	1,2	1,2	3	0	-	-	-
Rufous-sided Towhee	4,5	1-4	1-3	3	0	-	-	-
American Tree Sparrow	4,5	3,4	2	1	0	0	0	0
Chipping Sparrow	5,6	1,2	1,2	3	0	-	-	+
Brewer's Sparrow	5	1,2	1,2	3	0	0	-	-
Black-chinned Sparrow	5	1,2	1,2	1	0	0	-	-
Vesper Sparrow	5	1,2	1,2	2	0	0	-	-
Lark Sparrow	5	1,2	2	1	0	0	0	0
Black-throated Sparrow	5	1,2	1,2	1	0	0	-	-
Sage Sparrow	5	1-4	1-3	3	0	0	-	-
Lark Bunting	5	1,2	2	1	0	0	0	0
Savannah Sparrow	1,3,5	1-4	1-3	3	-	-	-	-
Fox Sparrow	5	1,2	1,2	3	0	0	-	-
Song Sparrow	3,4,5	1-4	1-3	3	0	-	-	-
Lincoln's Sparrow	4,5	1-4	1,3	1	0	-	-	-
Swamp Sparrow	3	1,3,4	2	1	0	-	0	0
White-throated Sparrow	4	3,4	2	1	0	0	0	0
Golden Crowned Sparrow	4,5	1,3,4	2	1	0	0	0	0

Species	Habitat	Season	Status	Peak Population Size	Responses to Disturbances A	B	C	D
White-crowned Sparrow	4,5	1-4	1-3	3	0	-	-	-
Harris' Sparrow	4	1,3,4	2	1	0	-	0	0
Dark-eyed Junco	4,5,6	1-4	1-3	3	0	-	-	-
McGown's Longspur	5	4	2	1	0	0	0	0
Lapland Longspur	2,5	3	2	1	0	0	0	0
Chestnut-collared Longspur	5	1,3	2	1	0	0	0	0
Bobolink	3,4	2	2	1	0	-	0	0
Red-winged Blackbird	3,4,5	1-4	1-3	3	0	-	-	+
Western Meadowlark	5	1-4	1-3	3	0	0	-	-
Yellow-headed Blackbird	1,3	1,2	1,2	2	-	-	0	0
Rusty Blackbird	1,3	3	2	1	0	0	0	0
Brewer's Blackbird	2,4,5	1-3	1,2	3	-	-	-	-
Great-tailed Grackle	3,5	2	2	1	0	0	0	0
Brown-headed Cowbird	3-6	1,2	1,2	2	0	-	-	+
Orchard Oriole	4	2	2	1	0	0	0	0
Hooded Oriole	4	1	2	1	0	0	0	0
Northern Oriole	4,5	1,2	1,2	2	0	-	-	0
Rosy Finch	5,7	1-4	1-3	2	0	0	-	-
Purple Finch	5,6	2,3	2	1	0	0	0	0
Cassin's Finch	4,5,6	1-4	1-3	3	0	-	-	-
House Finch	5	1,2	1,2	2	0	0	-	-
Red Crossbill	4,6	1-4	1-3	2	0	-	+	0
Pine Siskin	4,5	1-4	1-3	2	0	-	+	0
Lesser Goldfinch	4,5	1,2	1,2	2	0	-	-	-
Lawrence's Goldfinch	4	3	2	1	0	0	0	0
American Goldfinch	4,5	1,3,4	2	2	0	0	0	0
Evening Grosbeak	4,6	1-4	1-3	2	0	-	-	-
House Sparrow	2,4,5	1-4	1-3	3	0	-	-	+

Appendix B

Mammals of the Mono Basin and Their Ecological Characteristics

LEGEND: Habitat (1 = alpine/subalpine, 2 = Lodgepole forest, 3 = Jeffrey pine forest, 4 = Pinyon juniper, 5 = sagebrush-steppe, 6 = exposed lake bed and dunes, 7 = riparian, 8 = meadows and marshes); Response to Disturbances (A = lowering lake level, B = decreasing streamflow, C = increased grazing, D = increased frequency of fires), 0 = no change likely, + = increase, - = decrease. Distribution of bats in this region is poorly known.

			Response to Disturbances			
Common Name	Scientific Name	Habitat	A	B	C	D
Vagrant Shrew	*Sorex vagrans*	7,8	0	-	0	0
Dusky Shrew	*Sorex monticolus*	7,8	0	-	0	0
Water Shrew	*Sorex palustris*	7,8	-	-	0	0
Merriam's Shrew	*Sorex merriami*	6	0	0	0	0
Mt. Lyell Shrew	*Sorex lyelli*	7,8	-	-	0	0
Inyo Shrew	*Sorex tenellus*	7	0	-	0	0
Broad-handed Mole	*Scapanus latimanus*	7,8	0	-	0	0
Little Brown Bat	*Myotis lucifugus*					
Hairy-winged Myotis	*Myotis volans*					
Small-footed Myotis	*Myotis subulatus*					
Long-eared Myotis	*Myotis evotis*					

Common Name	Scientific Name	Habitat	A	B	C	D
			(Response to Disturbances)			
Western Pipistrelle	*Pipistrellus hesperus*					
Big Brown Bat	*Eptesicus fuscus*					
Hoary Bat	*Lasuirus cinereus*					
Brazilian Free-tailed Bat	*Tadarida brasiliensis*					
Pika	*Ochotona princeps*	1	0	0	0	0
White-tailed Hare	*Lepus townsendii*	1,2,3	0	0	-	0
Black-tailed Jackrabbit	*Lepus californicus*	5,6,8	0	0	0	0
Nuttall's Cottontail	*Sylvilagus nuttallii*	7,8,4,5	0	0	0	0
Pygmy Rabbit	*Brachylagus idahoenis*	5,7	0	-	-	0
Mountain Beaver	*Aplodontia rufa*	7,8	-	-	0	0
Yellow-tailed Marmot	*Marmota flaviventris*	1	0	0	0	0
Townsend Ground Squirrel	*Spermophilus townsendii*	5,6	0	0	0	0
Belding Ground Squirrel	*Spermophilus beldingi*	8	0	0	0	0
Beechey Ground Squirrel	*Spermophilus beecheyi*	8	0	0	+	+
Golden-mantled Ground Squirrel	*Spermophilus lateralis*	2,3,4,7	0	0	0	0
Alpine Chipmunk	*Eutamias alpinus*	1	0	0	0	0
Lodgepole Chipmunk	*Eutamias speciosus*	2,3	0	0	0	0
Yellow-pine Chipmunk	*Eutamias amoenus*	3,4	0	0	0	0
Panamint Chipmunk	*Eutamias panamintinus*	4,5	0	0	0	0
Least Chipmunk	*Eutamias minimus*	5,4,3	0	0	0	0

Common Name	Scientific Name	Habitat	Response to Disturbances			
			A	B	C	D
Chickaree	*Tamiasciurus douglasii*	2,3	0	0	0	0
Beaver	*Castor canadensis*	7	0	-	0	0
Northern Pocket Gopher	*Thomomys talpoides*	8,7	0	-	0	0
Great Basin Pocket Mouse	*Perognathus parvus*	4,5	0	0	0	0
Dark Kangaroo Mouse	*Microdipodops megacephalus*	6	0	0	0	0
Ord's Kangaroo Rat	*Dipodomys ordii*	6	0	0	0	0
Panamint Kangaroo Rat	*Dipodomys panamintinus*	5	0	0	0	0
Great Basin Kangaroo Rat	*Dipodomys microps*		0	0	0	0
Western Harvest Mouse	*Reithrodontomys megalotis*	8	-	-	0	0
Deer Mouse	*Peromyscus maniculatus*	all	0	0	0	0
Pinyon Mouse	*Peromyscus truei*	4	0	0	0	0
Northern Grasshopper Mouse	*Onychomys leucogaster*	5,6,4	0	0	0	0
Bushy-tailed Woodrat	*Neotoma cinerea*	most	0	0	0	0
Desert Woodrat	*Neotoma lepida*	most	0	0	0	0
Montane Vole	*Microtus montanus*	8	-	-	0	0
Long-tailed Vole	*Microtus longicaudus*	7,8	0	-	0	0
Sagebrush Vole	*Lagurus curtatus*	5,4,6	0	0	0	0
Heather Vole	*Phenacomys intermedius*	1	0	0	0	0

Common Name	Scientific Name	Habitat	Response to Disturbances			
			A	B	C	D
Muskrat	*Ondatra zibethica*	ponds	-	-	0	0
Western Jumping Mouse	*Zapus princeps*	7,8	-	0	0	0
Porcupine	*Erethizon dorsatum*	2,7	0	-	0	-
Coyote	*Canis latrans*	all	0	-	0	0
Red Fox	*Vulpes fulva*	2,1,8	0	0	-	0
Kit Fox	*Vulpes macrotis*	6,5	0	0	0	0
Gray Fox	*Urocyon cinereoargenteus*	4,7,	0	0	0	0
Black Bear	*Ursus americanus*	1,2	0	0	0	0
Raccoon	*Procyon lotor*	7	0	-	0	0
Long-tailed Weasel	*Mustela frenata*	all	0	0	0	0
Short-tailed Weasel	*Mustela erminea*	2,1,4	0	0	0	0
Mink	*Mustela vison*	7,8	-	-	0	0
Marten	*Martes americana*	2,1	0	0	0	0
Fisher	*Martes pennanti*	2,3	0	0	0	0
Wolverine	*Gulo luscus*	1,2	0	0	0	0
Badger	*Taxidea taxus*	8	0	0	0	0
Striped Skunk	*Mephitis mephitis*	7	0	-	0	0
Spotted Skunk	*Spilogale putorius*	rocky,7	0	0	0	0
Bobcat	*Lynx rufus*	4,5	0	-	0	0
Mountain Lion	*Felis concolor*	2,3,4	0	0	0	0
Mule Deer	*Odocoileus hemionus*	7,4,3,2	0	-	0	0
Pronghorn	*Antilocapra americana*	5,4	0	0	-	0
Bighorn Sheep	*Ovis canadensis*	1	0	0	-	0

Appendix C

Bibliography

HYDROLOGY

Reports

Blevins, M. L., and J. F. Mann, Jr. 1983. Mono Basin geology and hydrology. Presented at NWWA Western Regional Conference on Groundwater Management, San Diego, Oct. 26, 1983. 15 pp. plus tables and appendixes.

California Department of Water Resources. 1975. California's Groundwater. Department of Water Resources Bulletin 118.

California Department of Water Resources. 1979. Report of the Interagency Task Force on Mono Lake. 140 pp.

California Department of Water Resources. 1960. Reconnaissance Investigation of Water Resources of Mono and Owen Basins, Mono and Inyo Counties. 91 pp. plus tables and appendixes.

Cromwell, L., and J. D. Goodridge. 1979. Mono Lake, California, Water Balance. Unpublished report available in Water Resources Center Archives, University of California, Berkeley.

Gradek, P. D. 1983. An Inventory of Water Resources on Public Lands in the Mono Basin. Bureau of Land Management, Bakersfield District. 12 pp.

Harding, S. T. 1962. Unpublished notes. Available in the Water Resources Center Archives, University of California, Berkeley.

Harding, S. T. 1965. Recent Variations in the Water Supply of the Western Great Basin. Archives Series Report No. 16. Water Resources Center Archives, University of California, Berkeley.

Lee, C. Unpublished papers on Mono Lake and the Mono Basin. Available in the Water Resources Center Archives, University of California, Berkeley.

Lee, K. 1969. Infrared Exploration of Shoreline Springs: A Contribution to the Hydrology of Mono Basin, California. Ph.D. dissertation. Stanford University. 196 pp.

Lipinski, P. A. 1982. Discussion of geohydrologic data in Mono Basin, California. U.S. Geological Survey Open File Report. 28 pp.

Loeffler, R. M. 1977. Geology and hydrology. Pp. 6-38 in An Ecological Study of Mono Lake, California, D. W. Winkler, ed. Institute of Ecology Publication No. 12. University of California, Davis.

Los Angeles Department of Water and Power. 1984. Background Report on Mono Basin Geology and Hydrology. LADWP Aqueduct Division, Hydrology Section.

Los Angeles Department of Water and Power. 1984. Los Angeles' Mono Basin Water Supply. Briefing document. 13 pp. plus figures.

Los Angeles Department of Water and Power. 1987. Mono Basin Geology and Hydrology. LADWP Aqueduct Division, Hydrology Section.

Lynch, H. B. 1933. Report on Rainfall Fluctuations. Report to the Metropolitan Water District of Southern California. Report No. 589.

Lyon, R. J. P., and K. Lee. 1968. Infrared Exploration for Coastal and Shoreline Springs. Stanford University Rsl Technical Report No. 68-1. 68 pp. plus figures and tables.

Moe, R. A. 1972. Summary of the Hydrology of the Mono Basin. Prepared for the Sierra Club. 25 pp. plus figures.

Stine, S. 1981. A Reinterpretation of the 1857 Surface Elevation of Mono Lake. Report No. 52. California Water Resources Center, University of California, Berkeley.

Todd, D. K. 1984. The Hydrology of Mono Lake: A

Compilation of Data Developed for *State of California* v. *United States*, Civil No. S-80-696, U.S.D.C., E.D. California Consulting Engineers, Inc. Berkeley, Calif.

Vorster, P. 1984. A Water Balance Forecast Model for Mono Lake, California. Forest Service/USDA Region 5. Monograph No. 10. 350 pp.

Waugh, E. J. 1928. Report on Mono Lake Levels. Memorandum Report to A.B. West, Southern Sierras Power Company, June 19, 1928.

Data Compilations

Interagency Evapo-transpiration Advisory Service. Miscellaneous Agroclimatic Station Data. Pan Evaporation Record at Sprinkler Irrigated Station. Data sheets for climatic data (evaporation, wind, temperature), 1980-1983.

Los Angeles Department of Water and Power. 1932-present. Mono Basin Precipitation Records.

Los Angeles Department of Water and Power. 1934-present. Mono Basin Runoff Records.

Stine, S. 1985. Yearly (Oct. 1) Surface Elevations of Mono Lake, 1850-1911. One-page table. Prepared for *State of California* v. *U.S.*, civil no. S-80-696, U.S.D.C., E.D.

Stine, S. 1984. Surface Fluctuations of Mono Lake, 1850-present. Hydrograph of estimated and recorded lake elevations, two sheets.

U.S. Geological Survey. 1907-1986. Water Supply Papers for the Great Basin Drainage. 1907-08 to present.

Vorster, P. 1984. Mono Lake Hydrologic Fact Sheet. Prepared for Philip Williams and Associates. Includes summaries of Mono Lake levels, releases to Mono Lake, diversions and exports, aqueduct spills, precipitation, and hydrographs of lake levels.

PHYTOPLANKTON

Almanza, E., and J. M. Melack. 1985. Chlorophyll differences in Mono Lake (California) observable on Landsat imagery. Hydrobiologia 122:13-17.

Chapman, D. J. 1982. Investigations on the salinity tolerance of a diatom and green algae isolated from Mono Lake. Abstract from Mono Lake Symposium, Santa Barbara, Calif., May 5-7, 1982.

Jellison, R., and J. M. Melack. 1986. Nitrogen supply and primary production in hypersaline Mono Lake. Eos Trans. Am. Geophys. U. 67:974.

Jellison, R., and J. M. Melack. In press. Photosynthetic activity of phytoplankton and its relation to environmental factors in hypersaline Mono Lake, California. In J. M. Melack, ed., Saline Lakes. Developments in Hydrobiology, Dr W. Junk Publ., Dordrecht.

Lenz, P. H., S. D. Cooper, J. M. Melack, and D. W. Winkler. 1986. Spatial and temporal distribution patterns of three trophic levels in a saline lake. J. Plankton Res. 8(6):1051-1064.

Lovejoy, C., and G. Dana. 1977. Primary producer level. Pp. 42-57 in An Ecological Study of Mono Lake, California, D. W. Winkler, ed. Institute of Ecology, Publication No. 12. University of California, Davis.

Melack, J. M. 1985. The ecology of Mono Lake. National Geographic Society Research Reports. 1979 Projects. Pp. 461-470.

Melack, J., and R. Jellison. 1983. Final Report--Primary Productivity in Mono Lake, California. Available from LADWP.

Melack, J., and R. Jellison. 1985. Final Report--Limnology and Primary Productivity of Mono Lake, 1982-1984. Available from LADWP.

Melack, J., Y. Liang, and P. H. Lenz. 1982. Ecological responses of the plankton of Mono Lake to increased salinity. Abstract from Mono Lake Symposium, May 5-7, 1982.

Melack, J. M., R. S. Jellison, P. H. Lenz, and G. L. Dana. 1985. Responses of plankton in hypersaline Mono Lake to water diversions and El Niño. Abstracts of the 48th annual meeting of the American Society of Limnology and Oceanography, June 17-21.

INVERTEBRATE ZOOLOGY

Brine Fly

Dana, G., and D. B. Herbst. 1977. Secondary producer level. Pp. 57-63 in An Ecological Study of Mono Lake, California, D. W. Winkler, ed. Institute of Ecology Publication No. 12. University of California, Davis.

Herbst, D. B. 1977. Entomology. Pp. 70-87 in An Ecological Study of Mono Lake, California, D. W. Winkler, ed. Institute of Ecology Publication No. 12. University of California, Davis.

Herbst, D. B. 1980. Ecophysiology of the Larval Brine Fly. Master's thesis. Oregon State University.

Herbst, D. B. 1980. Ecological Physiology of an Alkaline Lake Insect, *Ephydra hians*. Report to Inyo National Forest. Mono Basin Research Group Contribution No. 5.

Herbst, D. B. 1982. Morphological variation in the alkali fly *Ephydra hians*. Abstract from Mono Lake Symposium, Santa Barbara, Calif., May 5-7, 1982.

Herbst, D. B. 1986. Comparative Studies of the Population Ecology and Life History Patterns of an Alkaline Salt Lake Insect: *Ephydra* (*Hydropyrus*) *hians* Say (Diptera: Ephydridae). Ph.D. dissertation. Oregon State University. 206 pp.

Herbst, D. B. In press. Comparative population ecology of *Ephydra hians* Say (Diptera:Ephydridae) at Mono Lake (California) and Abert Lake (Oregon).

Herbst, D. B., and G. Dana. 1977. Salinity tolerance in *Ephydra* and *Artemia*. Pp. 63-69 in An Ecological Study of Mono Lake, California, D. W. Winkler, ed. Institute of Ecology Publication No. 12. University of California, Davis.

Lenz, P. H., S. D. Cooper, J. M. Melack, and D. W. Winkler. 1986. Spatial and temporal distribution patterns of three trophic levels in a saline lake. J. Plankton Res. 8:1051-1064.

Brine Shrimp

Bowen, S. T. 1964. The genetics of *Artemia salina*. IV. Hybridization of wild populations with mutant stocks. Biol. Bull. 126:333-344.

Bowen, S. T., M. L. Davis, S. R. Fenster, and G. A. Lindwall. 1980. Sibling species of *Artemia*. Pp. 157-167 in The Brine Shrimp *Artemia*, Vol. I, Morphology, Genetics, Radiobiology, Toxicology. G. Personne, P. Sorgeloos, O. Roels, and E. Jaspers, eds. Wetteren, Belgium: Universa Press.

Bowen S. T., K. N. Hitchner, and G. L. Dana. 1984. *Artemia* speciation: Ecological isolation. Pp. 102-114 in Vernal Pools and Intermittent Streams, S. Jain and P. Moyle, eds. Institute of Ecology Publication No. 28. University of California, Davis.

Bowen, S. T., E. A. Fogarino, K. N. Hitchner, G. L. Dana, V. H. S. Chow, M. R. Buocristiani, and J. R. Carl. 1985. Ecological isolation in *Artemia*: Population differences in tolerance of anion concentrations. J. Crustacean Biol. 5(1):106-129.

Clark, L. S., and S. T. Bowen. 1976. The genetics of *Artemia salina*, Part 7, reproductive isolation. J. Hered. 67(6):385-388.

Conte, F. P. 1982. Brine Shrimp Salt Metabolism: Impact of Hypersalinity on Cellular Metabolism. Abstracts from Mono Lake Symposium, Santa Barbara, Calif., May 5-7, 1982.

Cooper, S. D., D. W. Winkler, and P. H. Lenz. 1984. The effect of grebe predation on a brine shrimp *Artemia monica* population. J. Anim. Ecol. 53(1):52-64.

Dana, G. L. 1982. Hatching in Mono Lake *Artemia* Cysts. Abstracts from Mono Lake Symposium, May 5-7, 1982.

Dana, G. L. 1983. Comparative Population Ecology of the Brine Shrimp *Artemia*. Master's thesis. San Francisco State University.

Dana, G., and D. B. Herbst. 1977. Secondary producer level. Pp. 57-63 in An Ecological Study of Mono Lake, California, D. W. Winkler, ed. Institute of Ecology Publication No. 12. University of California, Davis.

Dana, G. L., and P. H. Lenz. 1982. Spring Hatching Dynamics of *Artemia* in Mono Lake, California. Report to the Conservation Endowment Fund and Santa Clara Valley Audubon Society. 56 pp.

Dana, G. L., and P. H. Lenz. 1984. Ecology and Salinity Bioassay of *Artemia* in Mono Lake, California. A Preliminary Report to the David and Lucile Packard Foundation and the Inyo-Mono Fish and Game Advisory Commission.

Dana, G. L., and P. H. Lenz. 1985. Salinity bioassay of an *Artemia* population from Mono Lake, California. Abstracts of the 48th annual meeting of the American Society of Limnology and Oceanography, June 17-21, 1985, Minneapolis, Minn.

Dana, G. L., and P. H. Lenz. 1986. Effects of increasing salinity on an *Artemia* population from Mono Lake, California. Oecologia 68:428-436.

Dana, G., K. Hitchner, E. Fogarino, and S. T. Bowen. 1980. Mono Lake *Artemia*: Ecological isolation. In Abstracts of the 61st annual meeting of the American Association for the Advancement of Science, Pacific Division, Ecological Society of America, Western Div., June 22-27, 1980, University of California, Davis. A. E. Leviton, M. L. Aldrich, M. Berson, eds.

Dana, G. L., C. Foley, G. Starrett, W. Perry, and J. M. Melack. In press. In situ hatching rates of *Artemia monica* cysts in hypersaline Mono Lake. In J. M. Melack, ed., Saline Lakes. Developments in Hydrobiology, Dr W. Junk Publ.

Dircon Consultants Inc. 1982. Comparison of the Abundance and Spatial Distribution of Brine Shrimp in Tufa Shoals and Pelagic Regions of Mono Lake, California. Dircon Consultants Inc., Corvallis. 56 pp.

Drinkwater, L. E., and J. H. Crowe. 1984. Metabolic processes and release from dormancy in Mono Lake California USA *Artemia* cysts. Am. Zool. 24(3).

Drinkwater, L. E., and J. H. Crowe. 1986. Physiological Effects of Salinity on Dormancy and Hatching in Mono Lake *Artemia* Cysts. Prepared for Los Angeles Department of Water and Power.

Enzler, L., V. Smith, J. S. James, and H. S. Olcott. 1974. The lipids of Mono Lake, California, brine shrimp (*Artemia salina*). J. Agric. Food Chem. 22(2):330.

Herbst, D. B., and G. Dana. 1977. Salinity tolerance in *Ephydra* and *Artemia*. Pp. 63-69 in An Ecological Study of Mono Lake, California, D. W. Winkler, ed. Institute of Ecology Publication No. 12. University of California, Davis.

Herbst, D. B., and G. L. Dana. 1980. Physiological studies on a population of *Artemia* from Mono Lake, California. The Effects of Increasing Lake Water Salinity on Survival, Respiration Rate and Internal Solute Regulation. Pp. 157-167 in The Brine Shrimp *Artemia*. Vol. 2. Physiology, Biochemistry, Molecular Biology. G. Personne, P. Sorgeloos, O. Roels, and E. Jasper, eds. Universa Press, Wetteren, Belgium.

Jellison, R. S. 1985. Zooplankton mediated nitrogen and phytoplankton dynamics in a hypersaline lake. Abstracts of the 48th annual meeting of the American Society of Limnology and Oceanography, June 17-21, 1985, Minneapolis, Minn.

Lenz, P. H. 1980. *Artemia* in the Mono Lake ecosystem. Abstract for the Mono Lake Symposium at the annual AAAS meeting, Pacific Division, Davis, Calif., June 1980.

Lenz, P. H. 1980. Ecology of an alkali-adapted variety of *Artemia* from Mono Lake, California, USA. Pp. 80-96 in The Brine Shrimp *Artemia*. Vol. 3. Ecology, Culturing, Use in Aquaculture, G. Personne, P. Sorgeloos, O. Roels, and E. Jasper, eds. Universa Press, Wetteren, Belgium.

Lenz, P. H. 1982. Life history of *Artemia* in Mono Lake. Abstracts from Mono Lake Symposium, Santa Barbara, Calif., May 5-7, 1982.

Lenz, P. H. 1982. Population Studies on *Artemia* in Mono Lake, California. Ph.D. dissertation, University of California, Santa Barbara.

Lenz, P. H. 1984. Life-history analysis of an *Artemia* population in a changing environment. J. Plankton Res. 6(6):967-983.

Lenz, P. H. In press. Life-cycle studies in *Artemia*: A comparison between a subtropical and a permanent population. Proceedings of the 2nd International Symposium on the Brine Shrimp *Artemia*, Sept. 1-5, 1985, Antwerp, Belgium.

Lenz, P. H., and G. L. Dana. 1984. Large-scale patchiness in *Artemia* in Mono Lake (California), a hypersaline, terminal lake. Abstracts of the 47th annual meeting of the American Society of Limnology and Oceanography, June 11-14, 1984, Vancouver.

Lenz, P. H., S. D. Cooper, J. M. Melack, and D. W. Winkler. 1986. Spatial and temporal distribution patterns of three trophic levels in a saline lake. J. Plankton Res.

Los Angeles Department of Water and Power. 1983. 1983 *Artemia* Survey Data Report.

Los Angeles Department of Water and Power. 1984. 1984 *Artemia* Survey Data Report.

Mason, D. T. 1966. Density-current plumes. Science 152:354-365.

Melack, J. M. 1985. The Ecology of Mono Lake. National Geographic Society Research Reports. 1979 Projects. Pp. 461-470.

Melack, J. M., and P. H. Lenz. 1979. Plankton heterogeneity in an alkaline, saline lake (Mono Lake, California). Abstract for June 1979 meeting of American Society of Limnology and Oceanography, Long Island, N.Y.

Payne, T., and S. S. Kuwahara. 1972. Sterols of the brine shrimp *Artemia salina* from Mono Lake California. Experimentia 28(9):1022-1023.

Thun, M. A., and G. L. Starrett. In press. The Effect of cold, hydrated dormancy and salinity on the hatching of *Artemia* cysts from Mono Lake, California, U.S.A. Proceedings of the Second Annual International Symposium on The Brine Shrimp *Artemia*. Sept. 1-5, 1985, Antwerp, Belgium.

Thun, M. A., G. L. Schnoor, and F. P. Conte. 1984. Relative Population Abundance and Spatial Distribution of Brine Shrimp *Artemia* Located in Tufa Shoals and Deep Water Regions of Mono Lake, California. Dircon Consultants Inc., Corvallis. 33 pp. plus 12 figures.

Other

Schwan, T. G., and D. W. Winkler. 1984. Ticks parasitizing humans and California gulls at Mono Lake, California. Pp. 1193-1199 in Acarology VI, Vol. 2. D. A. Griffiths and C. E. Bowman, eds. Ellis Horwood Ltd., Chichester.

Schwan, T. G., J. J. Oprandy, and A. J. Main. 1985. Mono Lake virus associated with Argas ticks and California gulls on islands in Mono Lake, California. Presented at Thirty-fourth Annual Meeting of the American Society of Tropical Medicine and Hygiene, Miami, Fla., Nov. 1985.

Schwan, T. G., S. J. Brown, H. Hoogstroal, and M. Roshdy. In preparation. Argas, a new species of tick associated with California gulls on islands in Mono Lake, California: Description, life cycle and ecology. J. Parasitology.

Toft, C. A. 1983. Community patterns of nectivorous adult parasitoids (Diptera, Bombyliidae) on their resources. Oecologia 57:200-215.

Toft, C. A. 1984. Activity budgets in two species of bee flies (Lordotus: Bombyliidae Diptera): a comparison of species and sexes. Behav. Ecol. Sociobiol. 19:187-296.

VERTEBRATE ZOOLOGY

Birds

Chappell, M. A., D. L. Goldstein, and D. W. Winkler. 1984. Oxygen consumption, evaporative water loss, and temperature regulation of California gull chicks in a desert rookery. Physiol. Zool. 67(2):204-214.

Cooper, S. D., D. W. Winkler, and P. H. Lenz. 1984. The effect of grebe predation on a brine shrimp *Artemia monica* population. J. Anim. Ecol. 53(1):52-64.

Hart, T., and D. Gaines. 1983. Field Checklist of the Birds of the Mono Basin. Mono Lake Committee, Lee Vining.

Jehl, J. R. 1981. Mono Lake: A vital way station for the Wilson's phalarope. National Geographic 160:520-525.

Jehl, J. R. 1981. Mortality of Waterbirds at Mono Lake, California, 1981: Beached Bird Census. Hubbs Sea World Research Institute Technical Report No. 81-133.

Jehl, J. R. 1981. Post-fledging Mortality of California Gulls. Hubbs Sea World Research Institute Technical Report No. 81-135.

Jehl, J. R. 1982. Biology of Eared Grebes at Mono Lake, California. Hubbs Sea World Research Institute Technical Report No. 82-136.

Jehl, J. R. 1982. The biology of migrating Wilson's phalaropes. Abstract from Mono Lake Symposium, Santa Barbara, Calif., May 5-7, 1982.

Jehl, J. R. 1982. Biology of Northern Phalaropes at Mono Lake, California. Hubbs Sea World Research Institute Technical Report No. 82-146.

Jehl, J. R. 1982. Mortality of Waterbirds at Mono Lake, California, 1982: Beached Bird Census. Hubbs Sea World Research Institute Technical Report No. 82-149.

Jehl, J. R. 1982. Post-fledging Mortality of California Gulls. Hubbs Sea World Research Institute Technical Report No. 82-147.

Jehl, J. R. 1983. The Biology of Northern Phalaropes at Mono Lake, California. Hubbs Sea World Research Institute Technical Report No. 83-158.

Jehl, J. R. 1983. Breeding Success of California Gulls and Caspian Terns on the Paoha Islets, Mono Lake California, 1983. Hubbs Sea World Research Institute Technical Report No. 83-157.

Jehl, J. R. 1983. Comments on the Annual Census of California Gull Chicks at Mono Lake, with Special Reference to the Populations on Paoha Islets. Hubbs Sea World Research Institute Tech. Report 83-156.

Jehl, J. R. 1983. Mortality of eared grebes in winter of 1982-83. American Birds 37:832-835.

Jehl, J. R. 1983. Possible sexual differences in fledging mortality of California gulls and their implications for studies of feeding ecology. Colonial Waterbirds 6:218-220.

Jehl, J. R. 1983. Post-fledging Mortality of California Gulls, 1983. Hubbs Sea World Research Institute Technical Report No. 83-160.

Jehl, J. R. 1985. The Cooperative Gull Census of 1985: Comments and Recommendations. Hubbs Sea World Research Institute Technical Report No. 85-183.

Jehl, J. R. 1985. Leucism in the North American population of eared grebes. The Condor 87:439-441.

Jehl, J. R. In submission. Fall migration of northern phalaropes (*Phalaropus lobatus*) at the western edge of the Great Basin. Submitted to The Great Basin Naturalist.

Jehl, J. R., and D. R. Jehl. 1981. A North American record of the Asiatic marbled murrelet (*Brachhyramphus marmoratus* perdix). American Birds 36:91-92.

Jehl, J. R., and S. A. Mahoney. 1985. Physiological Ecology and Salt Loading of California Gulls at an Alkaline, Hypersaline Lake. Physiol. Zool. 58(5):553-563.

Jehl, J. R., D. E. Babb, and D. M. Powers. 1984. History of the California gull colony at Mono Lake. Colonial Waterbirds 7:94-104.

Lenz, P. H., S. D. Cooper, J. M. Melack, and D. W. Winkler. 1986. Spatial and temporal distribution patterns of three trophic levels in a saline lake. J. Plankton Res.

Mahoney, S. A., and J. R. Jehl. 1984. Body water content in marine birds. The Condor 86:208-209.

Mahoney, S. A., and J. R. Jehl. 1985. Adaptations of migratory shorebirds to highly saline and alkaline lakes: Wilson's phalarope and American avocet. The Condor 87:520-527.

Mahoney, S. A., and J. R. Jehl. 1985. Avoidance of salt-loading by a diving bird at a hypersaline and alkaline lake: eared grebe. The Condor 87:389-397.

Norris, V. N., and D. W. Winkler. In submission. Patterns in seasonal movements and mortality of California gulls. Submitted to Auk.

Page, G. W., L. E. Stenzel, D. W. Winkler, and C. W. Swarth. 1983. Spacing out at Mono Lake: Breeding Success, Nest Density and Predation in the Snowy Plover. Auk 100:13-24.

Paul, D. S. 1984. Current and Historical Breeding Status of the California Gull in the Great Salt Lake Region. Prepared for LADWP.

Power, D. M. 1980. The Biology of Certain Water and Shore Birds at Mono Lake, California. Prepared for LADWP.

Schwan, T. G., and D. W. Winkler. 1984. Ticks parasitizing humans and California gulls at Mono Lake, California. Pp. 1193-1199 in Acarology VI, Vol. 2. D. A. Griffiths and C. E. Bowman, eds. Ellis Horwood Ltd., Chichester.

Schwan, T. G., J. J. Oprandy, and A. J. Main. 1985. Mono Lake virus associated with Argas ticks and California gulls on islands in Mono Lake, California. Presented at

Thirty-fourth Annual Meeting of the American Society of Tropical Medicine and Hygiene, Miami, Fla., Nov. 1985.

Schwan, T. G., S. J. Brown, H. Hoogstroal, and M. Roshdy. In preparation. Argas, a new species of tick associated with California gulls on islands in Mono Lake, California: Description, life cycle and ecology. In preparation for J. Parasitology.

Shuford, W. D. 1985. Reproductive Success and Ecology of California Gulls at Mono Lake, California in 1985, with Special Reference to the Negit Islands: An Overview of Three Years of Research. Contribution No. 318 of the Point Reyes Bird Observatory.

Shuford, W. D., and G. W. Page. 1985. California Gull Decoy Placement on Negit Island, Mono Lake, California. Contribution No. 317 of the Point Reyes Bird Observatory.

Shuford, W. D., E. Strauss, and R. Hogan. 1984. Population Size and Breeding Success of California Gulls at Mono Lake, California in 1983. Contribution No. 126 of Point Reyes Bird Observatory.

Shuford, W. D., P. Super, and S. Johnston. 1985. Population Size and Breeding Success of California Gulls at Mono Lake, California in 1984. Contribution No. 294 of Point Reyes Bird Observatory.

Storer, R. W., and Jehl, J. R. 1985. Moult patterns and moult migration in the black-necked grebe (*Podiceps nigricollis*). Ornis Scandanivica 16(4):253-260.

Swarth, C. W. 1983. Foraging Ecology of Snowy Plovers and the Distribution of Their Arthropod Prey at Mono Lake, California. Master's thesis. California State University, Hayward.

Winkler, D. W. 1982. Importance of Great Basin Lakes in Northern California to Nongame Aquatic Birds, 1977. California Department of Fish and Game, Wildlife Management Branch, Admin. Rep. 82-4.

Winkler, D. W. 1982. Variation in California gull production at Mono Lake: The effects of predators, food and weather. Abstract from Mono Lake Symposium, May 5-7, 1982.

Winkler, D. W. 1983. California Gull Nesting at Mono Lake, California in 1982: Chick Production and Breeding

Biology. Final Report for Contract 98210-0894. Submitted to U.S. Fish and Wildlife Service. Arcata Field Station.

Winkler, D. W. 1983. Ecological and Behavioral Determinants of Clutch Size: The California Gull in the Great Basin. Ph.D. dissertation. University of California, Berkeley.

Winkler, D. W. 1985. Factors determining a clutch size reduction in the California gull: A multi-hypothesis approach. Evolution 39:667-677.

Winkler, D. W., and S. D. Cooper. 1985. The ecology of migrant eared grebes at Mono Lake, California. Ibis.

Winkler, D. W., and D. Shuford. In submission. History of the California gull colony at Mono Lake, California: An alternative interpretation. Submitted to Colonial Waterbirds.

Winkler, D. W., C. P. Weigen, F. B. Engstrom, and S. E. Burch. 1977. Ornithology. Pp. 88-113 in An Ecological Study of Mono Lake, California, D. W. Winkler, ed. Institute of Ecology Publication No. 12. University of California, Davis.

Zink, R. M., and D. W. Winkler. 1983. Genetic and morphological similarity of 2 California gull *Larus californicus* populations with different life history traits. Biochem. Syst. Ecol. 11(4):397-404.

Mammals

Harris, J. H. 1982. Mammals of the Mono Lake-Tioga Pass Region. Kutsavi Books, Lee Vining. 55 pp.

Harris, J. H. 1984. An experimental analysis of desert rodent foraging ecology. Ecology 65(5):1579-1584.

Harris, J. H. In press. Microhabitat segregation in two desert rodent species: The effect of prey availability on diet. J. Ecologia.

BOTANY/VEGETATION

Brotherson, J. D., and S. R. Rushforth. 1985. Invasion and stabilization of recent beaches by saltgrass (*Distichlis spicata*) at Mono Lake, Mono County, California. Great Basin Naturalist 45:542-545.

Burch, J. B., J. Robbins, and T. Wainwright. 1977. Botany. Pp. 114-142 in An Ecological Study of Mono Lake, California, D. W. Winkler, ed. University of California, Davis, Institute of Ecology Publication No. 12.

Dummer, K., and R. Colwell. 1985. Vegetation Map of the Historical Shorelands of Mono Lake, California. Scale 1:24,000. Prepared from 1982 aerial photos. Prepared by the State of California for *State of California* v. *U.S.*, civil no. S-80-696, U.S.D.C., E.D. Calif.

Murata, T., and C. Ishikawa. 1981. Chemical physiochemical and spectrophotometric properties of crystalline chlorophyll protein complexes from *Lepidium virginicum*. Biochim. Biophys. Acta 635(2):341-347.

Stine, S., D. Gaines, and P. Vorster. 1984. Destruction of riparian systems due to water development in the Mono Lake watershed. Pp. 528-533 in California Riparian Systems: Ecology, Conservation, and Productive Management, R. D. Warner and C. Hendricks, eds. University of California Press.

Taylor, D. W. 1981. Plant Checklist for the Mono Basin, California. Report to Inyo National Forest. Mono Basin Research Group Contribution No. 3.

Taylor, D. W. 1982. Riparian Vegetation of the Eastern Sierra: Ecological Effects of Stream Diversions. Report to Inyo National Forest. Mono Basin Research Group Contribution No. 6.

WATER CHEMISTRY/BIOGEOCHEMISTRY

Lake

Andersen, R. F., M. P. Bacon, and P. G. Brewer. 1982. Elevated concentrations of actinides in Mono Lake. Science 216(4545):514-516.

Cleveland, J. M., T. F. Rees, and K. L. Nash. 1983. Plutonium speciation in water from Mono Lake, California. Science 222(4630):1323-1325.

Dana, G., D. B. Herbst, C. Lovejoy, B. Loeffler, and K. Otsaki. 1977. Physical and chemical limnology. Pp.

40-42 in An Ecological Study of Mono Lake, California, D. W. Winkler, ed. Institute of Ecology Publication No. 12. University of California, Davis.

Mason, D. T. 1966. Limnology of Mono Lake, California. Ph.D. dissertation. University of California, Davis. 221 pp.

Mason, D. T. 1967. Limnology of Mono Lake, California. University of California Publications in Zoology 83. 110 pp.

Miller, L. G., and R. S. Oremland. 1985. Methane emission rates in lakes. Eos 66(51).

Miller, L., C. Culbertson, R. Harvey, and R. S. Oremland. 1984. Methane and Meromixis in Mono Lake: An alkaline hypersaline environment. Eos 65(45).

Peng, T. H., and W. S. Broecker. 1980. Gas exchange rates for three closed-basin lakes. Limnol. Oceanogr. 25(5):789-796.

Reed, W. E. 1977. Biogeochemistry of Mono Lake, California. Geochim. Cosmochim. Acta 41(9):1231-1245.

Simpson, H. J., R. M. Trier, C. R. Olsen, and D. E. Hammond. 1979. Plutonium mobility in Mono Lake, California. Eos, 1979 Spring Annual Meeting of American Geophysical Union.

Simpson, H. J., R. M. Trier, C. R. Olsen, D. E. Hammond, A. Ege, L. Miller, and J. M. Melack. 1980. Fallout plutonium in an alkaline, saline lake. Science 207(4435):1071-1073.

Simpson, H. J., R. M. Trier, J. R. Toggweiler, G. Mathieu, B. L. Deck, C. R. Olsen, D. E. Hammond, C. Fuller, and T. L. Ku. 1982. Radionuclides in Mono Lake, California. Science 216(4545):512-514.

Steel, G., and W. Henderson. 1972. Rapid method for detection and characterization of steroids. Analytical Chem. 44(7):1302-1304.

Thurber, D. L., and W. S. Broecker. 1970. The behavior of radiocarbon in the surface waters of the Great Basin. Pp. 379-400 in Radiocarbon Variations and Absolute Chronology, I. V. Olsson, ed. Nobel Symposium No. 12. Wiley and Sons, New York.

Springs

Lee, K. 1969. Infrared Exploration of Shoreline Springs: A Contribution to the Hydrology of Mono Basin, California. Ph.D. dissertation. Stanford University. 196 pp.

Mariner, R. H., T. S. Presser, and W. C. Evans. 1977. Hot Springs of the Central Sierra Nevada, California. USGS Open File Report 77-559. 27 pp.

AIR QUALITY

Barone, J. B., L. L. Ashbaugh, B. H. Kusko, and T. A. Cahill. 1980. The Effect of Owens Dry Lake on Air Quality in the Owens Valley with Implications for the Mono Lake Area. American Chemical Society Symposium Series 1981(167):327-346.

Brattain, R. R. 1982. Characterization of the ambient particulate matter in the Mono Lake basin. Abstract from Mono Lake Symposium, Santa Barbara, Calif., May 5-7, 1982.

China Lake Naval Weapons Center. 1980. Desertification in Owens Valley and Mono Basin of California. Preliminary copy.

Crocker Nuclear Laboratory. 1979. A Study of Ambient Aerosols in the Owens Valley Area. Prepared by the Air Quality Group, Crocker Nuclear Laboratory, University of California, Davis.

Crocker Nuclear Laboratory. 1981. The Effect of Mono Lake on the Air Quality in the Mono Lake Region. Prepared by the Air Quality Group, Crocker Nuclear Laboratory, University of California, Davis.

Crocker Nuclear Laboratory. 1984. Study of Particle Episodes at Mono Lake. Prepared by the Air Quality Group, Crocker Nuclear Laboratory, University of California, Davis

Kusko, B. H., L. L. Ashbaugh, and T. A. Cahill. 1982. The Effect of Mono Lake on the Air Quality in the Mono Lake Region. Abstract from Mono Lake Symposium, Santa Barbara, Calif., May 5-7, 1982.

Kusko, B. H., J. B. Barone, and T. A. Cahill. 1983. The Effect of Mono Lake on the Air Quality in the Mono Lake Region. California Air Resources Board Report. Sept. 1983(163).

Los Angeles Department of Water and Power. In progress. Summary of Daily Mono Lake Air Quality Photos. Available from LADWP.

NUS Corporation. 1981. Aerosol Sampling Program at Lee Vining, California. Prepared for LADWP.

St. Amand, P. C., A. Matthews, C. Gaines, and R. Reinking. 1986. Dust Storms from Owens and Mono Valleys, California. Naval Weapons Center, China Lake, Calif. 76 pp.

Trusdeil Laboratories Inc. 1981. Mono Lake Basin Soil Samples Study. Report No. 33656, prepared for LADWP.

GEOLOGY

Al Rawi, Y. T. 1969. Cenozoic History of the Northern Part of Mono Basin (Mono County), California and Nevada. Doctoral dissertation. University of California, Berkeley. 188 pp.

Axtel, L. 1972. Mono Lake wells abandoned. Calif. Geol. 25(3):66-67.

Bailey, R. A. 1980. Structural and petrologic evolution of the Long Valley, Mono Craters, and Mono Lake volcanic complexes, eastern California. Eos Trans. Amer. Geophys. Union 61(46).

Bailey, R. A. 1982. Other Potential Eruption Centers in California: Long Valley-Mono Lake, Coso, and Clear Lake Volcano Fields. Special Publication, California Division of Mines and Geology 63:17-28.

Chesterman, C. W., and C. H. Gray. 1966. Geology and structure of the Mono Basin, Mono County, California. Pp. 11-18 in Guidebook Along the East-Central Front of the Sierra Nevada. Geological Society of Sacramento, Annual Field Trip.

Christensen, M. N., and C. M. Gilbert. 1964. Basaltic cone suggests constructional origin of some guyots. Science 143(3603):240-242.

Christensen, M. N., C. M. Gilbert, K. R. Lajoie, and Y. T. Al-Rawi. 1969. Geological-geophysical interpretation of Mono Basin, California-Nevada. J. Geophys. Res. 74(22):5221-5239.

Cloos, E. 1931. Mechanism of the intrusion of the granite masses between Mono Lake and the mother lode. Geol. Soc. Amer. Bull. 43(1):236.

Cloud, P., and K. R. Lajoie. 1980. Calcite-impregnated defluidization structures in littoral sands of Mono Lake, California. Science 210(4473):1009-1012.

Cohen, L. H., and P. H. Ribbe. 1966. Magnesium phosphate mineral replacement at Mono Lake, California. Amer. Mineral. 51(11-12):1755-1765.

Cooper, J. G., Jr., and G. E. Dunn. 1969. Struvite found at Mono Lake. California Division of Mines and Geology. Mineral Information Service 22(3):44-45.

Custer, S. G. 1973. Stratigraphy and Sedimentation of Black Point Volcano. Master's thesis. University of California, Berkeley.

Denham, C. R. 1971. Eastward drift of the geomagnetic field 25,000 years ago. Eos Trans. Amer. Geophys. Union 52(11):822.

Denham, C. R. 1974. Counter-clockwise motion of paleomagnetic directions 24000 years ago at Mono Lake, California. J. Geomagn. Geoelec. 25(5):487-498.

Denham, C. R., and A. Cox. 1971. Evidence that the Laschamp polarity event did not occur 13,000-30,400 years ago. Earth Planet. Sci. Lett. 13(1):181-190.

Dunn, J. R. 1951. Geology of the Western Mono Lake Area. Ph.D. dissertation. University of California, Berkeley.

Dunn, J. R. 1953. The origin of the deposits of tufa in Mono Lake. J. Sediment. Petrol. 23(1):18-23.

Friedman, I. 1968. Hydration rind dates rhyolite flows. Science 159(3817):878-880.

Friedman, J. D. 1968. Thermal anomalies and geologic features of the Mono Lake Area, California, as revealed by infrared imagery. Pp. 1-22 in Earth Resources Aircraft Program, Status Review, Vol. 1, Sect. 11. National Aeronautics and Space Administration.

Gilbert, C. M. 1938. Welded tuff in eastern California. Geol. Soc. Amer. Bull. 49(12) pt. 1:1829-1862.

Gilbert, C. M., M. N. Christensen, Y. Al-Rawi, and K. R. Lajoie. 1968. Structural and volcanic history of Mono Basin, California-Nevada. Geol. Soc. Amer. Mem. 116. pp. 275-329.

Gilbert, C. M., K. R. Lajoie, and M. N. Christensen. 1969. Low-density sedimentary fill in Mono Basin, California. Geol. Soc. Amer. Abstr. 1969, Part 3 (Cordilleran Sect.). p. 18.

Gillespie, A. R. 1982. Quaternary Glaciation and Tectonism in the Southeastern Sierra Nevada, Inyo County, California. Ph.D. dissertation. California Institute of Technology.

Groesbeck, M. J. 1940. Minerals of Mono Lake Basin. Mineralogist 8(4):123-124, 203-204.

Hanna, G. D. 1963. Some Pleistocene and Pliocene Freshwater Mollusca from California and Oregon. California Academy of Science Occasional Paper 43. 23 pp.

Henderson, W., W. E. Reed, G. Steel, and M. Calvin. 1971. Isolation and identification of sterols from a Pleistocene sediment. Nature 231(5301):308-309.

Henderson, W., W. E. Reed, G. Steel, and M. Calvin. 1971. The Origin and Incorporation of Organic Molecules in Sediments as Elucidated by Studies of the Sedimentary Sequence from a Residual Pleistocene Lake. International Meeting Organic Geochemistry Program Abstract No. 5.

Henyey, T. L., J. C. Liddicoat, R. Day, R. Dodson, M. Fuller, and D. F. Palmer. 1977. Ages of Lake Tahoe cores and comparison of paleomagnetic curves for Lake Tahoe with those of Mono Lake and Lake Lahontan. Eos Trans. Amer. Geophys. Union 58(12):1124.

Hermance, J. F. 1983. The Long Valley/Mono Basin volcanic complex in eastern California: Status of present knowledge and future research needs. Rev. Geophy. Space Phys. 21(7):1545-1565.

Hermance, J. F., W. M. Slocum, and G. A. Neumann. 1984. The Long Valley/Mono Basin volcanic complex: A preliminary magnetolurgic and magnetic variation interpretation. J. Geophys. Res. B89(10):8325-8337.

Holmes, R. W. 1980. Late Pleistocene limnology of Mono Lake-Lake Russell. Abstracts, American Association for the Advancement of Science, Pacific Division, 61st Annual Meeting.

Jehl, J. R. 1983. Tufa formation at Mono Lake, California, Mono County. Calif. Geol. 36(1)3 pp.

Johnson, L. R. 1965. Crustal structure between Lake Mead, Nevada and Mono Lake, California. J. Geophys. Res. 70(12):2863-2872.

Kesseli, J. E. 1948. Correlation of Pleistocene Lake Terraces and Moraines at Mono Lake, California. Geol. Soc. Amer. Bull. 59(12) pt. 2:1375.

Kilbourne, R. T., C. W. Chesterman, and S. H. Wood. 1980. Recent volcanism in the Mono Basin-Long Valley region of Mono County, California. Pp. 7-22 in Mammoth Lakes, California Earthquakes of May 1980, R. W. Sherburne, ed. Special Report, California Division of Mines and Geology 150.

Komar, P. D. 1973. Observations of beach cusps at Mono Lake, California. Geol. Soc. Amer. Bull. 84(11):3593-3600.

Komar, P. D. 1973. Occurrence of brick pattern oscillatory ripple marks at Mono Lake California. J. Sediment. Petrol. 43(4):1111-1113.

Lajoie, K. R. 1968. Quaternary Stratigraphy and Geologic History of Mono Basin, Eastern California. Ph.D. dissertation. University of California, Berkeley.

Lajoie, K. R., J. C. Liddicoat, and S. W. Robinson. 1980. Refinement of the chronology and paleomagnetic record at Mono Lake, California. Eos Trans. Amer. Geophys. Union 61(17).

Lajoie, K. R., S. W. Robinson, R. M. Forester, and J. P. Bradbury. 1982. Rapid climatic cycles recorded in closed-basin lakes. Presented at Seventh Bienniel Conference of the American Quaternary Association: Character and Timing of Rapid Environmental and Climatic Changes. Abstract 7. p. 53.

Liddicoat, J. C. 1977. Paleomagnetic curves for Lake Lahontan, Lake Bonneville, and Mono Lake deposits: Usefulness in Quaternary geology in Great Basin. Abstract from International Union of Quaternary Research Congress, Birmingham, U.K., Aug. 15-24, 1977.

Liddicoat, J. C. 1980. Synchroneity of Pleistocene Lake Sediments Examined Paleomagnetically. Presented at

American Quaternary Association Sixth Bienniel Meeting, Orono, Aug. 18-20, 1980. Abstract 6.

Liddicoat, J. C., and R. S. Coe. 1975. Geologic Applications for Late Pleistocene Great Basin Paleomagnetic Curves. Presented at Geological Society of America, Cordilleran Section, 71st Annual Meeting, Los Angeles, March 25-27, 1975. Abstract Programs 7(3):341.

Liddicoat, J. C., and R. S. Coe. 1975. Mono Lake 24,000 year BP geomagnetic excursion--Additional data. Eos Trans. Amer. Geophys. Union 56(12):978.

Liddicoat, J. C., and R. S. Coe. 1979. Mono Lake geomagnetic excursion. J. Geophys. Res. 84(B1):261-271.

Liddicoat, J. C., and S. P. Lund. 1983. A high resolution of secular variation from Quaternary sediments from Mono Lake, California. Eos Trans. Amer. Geophys. Union 64(45):685.

Liddicoat, J. C., K. R. Lajoie, R. A. Bailey, A. M. Sarna-Wojcicki, P. C. Russell, and M. Woodward. 1980. Reversal of the Paleomagnetic field in Brunehesage lacustrine sediments in Long Valley and Mono Basin, California. Eos Trans. Amer. Geophys. Union 61(1).

Liddicoat, J. C., K. R. Lajoie, and A. M. Sarna-Wojcicki. 1982. Detection and dating of the Mono Lake excursion in the Lake Lahontan Sehoo Formation, Carson Sink, Nevada. Eos Trans. Amer. Geophys. Union. Abstract No. GP82A-05.

Loeffler, R. M. 1977. Geology and hydrology. Pp. 6-38 in An Ecological Study of Mono Lake, California, D. W. Winkler, ed. Institute of Ecology Publication No. 12. University of California, Davis.

Lyon, R. J. P., and G. I. Ballew. 1976. Macro-linear Elements in the Greater Mono Basin, California. Utah Geologic Association Publication 5, Proceedings of 1st International Conference on New Basement Tectonics.

Lyon, R. J. P., and G. I. Ballew. 1976. Macro-linear Elements in the Greater Mono Basin, California. Utah Geologic Association Publication 5, Proceedings of 1st International Conference on New Basement Tectonics. 139 pp.

Martin, R. C., and E. E. Welday. 1977. Lake bottom thermal gradient survey at Clear Lake and Mono Lake,

California. Presented at Geothermal Resources Council Annual Meeting, May 9-11, 1977, San Diego, Calif.

Mavko, B. B., and G. A. Thompson. 1983. Crustal and upper mantle structure of the northern and central Sierra Nevada. J. Geophys. Res. B88:7:5874-5892.

Miller, C. D., D. R. Mullineaux, D. R. Crandell, and R. A. Bailey. 1982. Potential Hazards from Future Volcanic Eruptions in the Long Valley-Mono Lake Area, East-central California and Southwest Nevada: A Preliminary Assessment. Geological Survey Circular.

Mills, J. M., F. E. Followill, and C. Broadwater. 1982. Microearthquakes and structure in the Mono Craters area, Mono County, California. Eos Trans. Amer. Geophys. Union 63:45.

Moniot, R. K., T. Milazzo, T. Kruse, W. Savin, and G. F. Herzog. 1981. ^{10}Be in strata from Mono Lake, California by accelerator-based mass spectrometry. Eos Trans. Amer. Geophys. Union 62:45.

Negrini, R., J. O. Davis, and K. L. Verosub. 1983. The Mono Lake excursion as recorded by lacustrine sediments from Summer Lake, Oregon. Eos Trans. Am. Geophys. Union 64(45):685.

Negrini, R. M., Davis, J. O., and K. L. Verosub. 1984. Mono Lake geomagnetic excursion found at Summer Lake, Oregon. Geology (Boulder) 12(11):643-646.

Oldson, J., P. Wilde, E. Welday, and R. Martin. 1977. In situ thermal conductivity measurements for heat flow determinations in marine and lacustrine environments. Eos Trans. Amer. Geophys. Union 58:6.

Pakiser, L. C. 1968. Seismic evidence for the thickness of Cenozoic deposits in Mono Basin, California and Nevada. Geol. Soc. Amer. Bull. 79(12):1833-1838.

Pakiser, L. C. 1970. Structure of Mono Basin, California. J. Geophys. Res. 75(20):4077-4080.

Pakiser, L. C. 1976. Seismic exploration of Mono Basin, California. J. Geophys. Res. 81(20):3607-3618.

Pakiser, L. C., F. Press, and M. F. Kane. 1960. Physical investigation of Mono Basin, California. Geol. Soc. Amer. Bull. 71(4):415-447.

Pelagos Corp. 1987. A Bathymetric and Geologic Survey of Mono Lake, California. Report prepared for Los

Angeles Department of Water and Power. San Diego, Calif.

Pitt, A. M., and D. W. Steeples. 1975. Microearthquakes in Mono Lake Northern Owen Valley, California Region from September 28 to October 18, 1970. Bull. Seismol. Soc. Amer. 65(4):835-844.

Putnam, W. C. 1950. Moraine and shoreline relationships at Mono Lake, California. Geol. Soc. Amer. Bull. 61:115-122.

Putnam, W. C. 1984. Quaternary geology of the June Lake District. Geol. Soc. Amer. Bull. 80:1281-1302.

Ribbe, P. H., and L. Cohen. 1966. Newberyite and Monetite from Paoha Island, Mono Lake. California Division of Mines and Geology Mineral Information Service 19(3).

Russell, I. C. 1889. Quaternary History of the Mono Valley, California. U.S. Geological Survey Eighth Annual Report. Reprinted from the Eighth Annual Report of the United States Geological Survey. Artemia Press, Lee Vining, Calif. Pp. 267-394.

Scholl, D. W. 1966. Mono Lake tufa pinnacles. In Guidebook Along the East-central Front of the Sierra Nevada. Geologic Society of Sacramento Annual Field Trip.

Scholl, D. W., and W. H. Taft. 1964. Algae, Contributors to the Formation of Calcareous Tufa, Mono Lake, California. J. Sedimentary Petrol. 34(2):309-319.

Scholl, D. W., and W. H. Taft. 1965. Deposition, Mineralogy, and C-14 Dating of Tufa, Mono Lake, California. Geol. Soc. Amer. Spec. Paper 82.

Scholl, D. W., R. von Huene, and P. Saint-Amand. 1966. Geology of Mono Lake. In Guidebook Along the East-central Front of the Sierra Nevada. Geological Society of Sacramento Annual Field Trip.

Scholl, D. W., R. von Huene, P. Saint-Amand, and J. B. Ridlon. 1967. Age and origin of topography beneath Mono Lake, a remnant Pleistocene lake, California. Geol. Soc. Amer. Bull. 78:583-600.

Sklarew, D. S. 1979. Analysis and simulated diagenisis of kerogen in a recent bottom mud from Mono Lake, California: A comparison with selected ancient kerogens. Geochim. Cosmochim. Acta 43(12):1949-1958.

Steeples, D. W., and A. M. Pitt. 1973. Microearthquakes in and near Long Valley, California. Eos Trans. Amer. Geophys. Union 54(11):1213.

Stine, S., S. Wood, K. Sieh, and C. D. Miller. 1984. Holocene Paleoclimatology and Tephrochronology East and West of the Central Sierran Crest. Friends of the Pleistocene, Pacific Cell, Field Trip Guidebook.

Toppozada, T. R., J. H. Bennett, and C. H. Cramer. 1981. Earthquakes and water levels at Mono Lake, Mono County, California. Presented at Seismological Society of America 76th Annual Meeting, March 23-25, 1981, Berkeley, Calif.

Toste, A. P., and M. Calvin. 1978. Environmental dependence of short-term sterol diagenesis in contemporary sediments. Presented at Geological Association of Canada, Mineralogical Association of Canada, Geological Society of America 1978 Joint Annual Meeting. Abstract Programs 10:7.

Toste, A. P., and M. Calvin. 1979. Pathway of short-term sterol diagenesis in a contemporary sediment. Geological Society of America 92nd Annual Meeting. Abstract Programs 11:7.

Toste, A. P., G. Steel, R. P. Philp, and M. Calvin. 1974. Incorporation and microbial diagenesis of sterols in recent sediments. Geol. Soc. Amer. Abst. Progr. 6:7:989-990.

Verosub, K. L. 1976. Mono Lake excursion as seen from Clear Lake, California. Eos Trans. Amer. Geophys. Union 57(12):909.

Vetter, U. R., and A. S. Ryall. 1983. Systematic change of focal mechanism with depth in the Western Great Basin. J. Geophys. Res. B88:10:8237-8250.

Whitcomb, J. H., and J. B. Rundle. 1983. Gravity variation in the Mammoth Lakes and Owens Valley, California regions. Seismological Society of America abstracts, 78th Annual Meeting.

Wood, S. H. 1977. Distribution, correlation, and radiocarbon dating of late Holocene Tephra, Mono and Inyo Craters, eastern California. Geol. Soc. Amer. Bull. 88(1):89-95.

Wood, S. H. 1983. Chronology of Late Pleistocene and Holocene Volcanics, Long Valley and Mono Basin Geothermal Areas, Eastern California. U.S. Geol. Surv. Open File Rep. 84 pp.

Wornardt, W. W., Jr. 1964. Pleistocene Diatoms from Mono and Panamint Lake Basins, California. California Academy of Science Occasional Paper 46. 27 pp.

ECOSYSTEM OVERVIEWS

Eugster, H. P., and L. A. Hardie. 1978. Saline lakes. Pp. 237-293 in Lakes--Chemistry, Geology, Physics, A. Lerman, ed. Springer-Verlag, New York.

Gaines, D. 1981. Mono Lake Guidebook. Kutsavi Books, Lee Vining. 114 pp.

Hammer, U. T. 1986. Saline lake ecosystems of the world. Dr W. Junk Publ., Dordrecht. 616 pp.

Mason, D. T. 1967. Limnology of Mono Lake, California. University of California Publications in Zoology 83. 110 pp.

Melack, J. M. 1982. Mono Lake: An Ecosystem in Transition? Symposium Workshop, Santa Barbara, Calif., May 5-7, 1982. Summary of Panel Discussions.

Melack, J. M. 1983. Large, deep, salt lakes: A comparative limnological analysis. Hydrobiologia 105:223-230.

Melack, J. M. 1985. The Ecology of Mono Lake. National Geographic Society Research Reports. 1979 Projects. Pp. 461-470.

Thun, M. A., and G. L. Starrett. 1982. Mono Lake: Analysis of a saline ecosystem. Abstract from Mono Lake Symposium, Santa Barbara, Calif., May 5-7, 1982.

U.S. Bureau of Land Management. 1984. Environmental Assessment: Designation of Public Lands in the Mono Lake Ecological Area as an Area of Critical Environmental Concern. U.S. Department of the Interior, Bureau of Land Management, Bakersfield District, Calif.

Winkler, D., ed. 1977. An Ecological Study of Mono Lake, California. Institute of Ecology Publication No. 12. University of California, Davis.

MAPS/PHOTOS

Bathymetric Maps

Johnson, W. D., and I. C. Russell. Lake Mono. Shaded relief and bathymetry map. Scale 1:125,000. U.S. Geological Survey. Eighth Annual Report Plate 19.

Los Angeles Department of Water and Power. 1980. Bathymetry near Negit Island, Mono Lake. Map at scale of 1 inch = 500 feet.

Pelagos Corporation. 1987. A Bathymetric and Geologic Survey of Mono Lake, California. Report prepared for Los Angeles Department of Water and Power. San Diego, Calif.

Scholl, D. W., R. von Huene, P. Saint Amand, and J. B. Ridlon. 1967. Age and origin of topography beneath Mono Lake, a remnant Pleistocene lake, California. Geol. Soc. Amer. Bull. 78:583-600.

Topographic Maps

Pacific Western Aerial Surveys. 1985. Mono Lake Topographic Maps. Topo maps of near-shore zone, 1 inch = 500 feet, five-foot contours, twelve sheets. Prepared for *State of California* v. *U.S.*, Civil No. S-80-696, U.S.D.C., E.D. Calif.

U.S. Geological Survey. 1987. 7.5-minute topographic quadrangles: Bodie SE, Cowtrack Mountain NW, Mono Craters NE, Trench Canyon SW. USGS.

U.S. Geological Survey. 1987. 15-minute topographic quadrangles: Bodie, Cowtrack Mountain, Matterhorn Peak, Mono Craters, Trench Canyon, Tuolumne Meadows. USGS.

Aerial Photos

Los Angeles Department of Water and Power. 1980. Mono Lake. Color air photo dated March 28, 1980 with locations of springs, wells, test holes, and investigation routes, in Background Report on Mono Basin Geology and Hydrology. LADWP.

U.S. Department of Agriculture. Forest Service. 1982. Aerial photographs of Mono Lake and vicinity, Oct. 1, 1982. Color infrared images, approximate scale 1:17,360.

U.S. Geological Survey. 1982. Aerial photographs of Mono Lake and vicinity, July-Aug. 1982. Project GS-VFEF-C. Stereo color diapositives and contact prints, black-and-white contact prints with ground control information. Flight height: 16,500 feet. Approximately scale 1:33,000.

Remote Sensing

Almanza, E., and J. M. Melack. 1985. Chlorophyll differences in Mono Lake (California) observable on Landsat imagery. Hydrobiologia 122:13-17.

Friedman, J. D. 1968. Thermal anomalies and geologic features of Mono Lake area, California, as revealed by infrared imagery. Pp. 1-22 in Earth Resources Aircraft Program, Status Review, Vol. I, Sect. 11, National Aeronautics and Space Administration.

Lee, K. 1970. Infrared exploration for shoreline springs at Mono Lake, California, test site. Proceedings 6th International Symposium on Remote Sensing of the Environment. University of Michigan, Ann Arbor, Oct. 13-16, 1969. Vol. 2. Pp. 1075-1100.

Lyon, R. J. P., and K. Lee. 1968. Infrared Exploration for Coastal and Shoreline Springs. Stanford University Rsl Technical Report No. 68-1. 68 pp. plus figures.

Other

United States Department of Agriculture. Forest Service. 1979. Inyo National Forest. Scale: 1/2 inch = 1 mile. USDA.

MISCELLANEOUS

Bertram, J. R., and A. R. Kriebel. 1968. Wave Runup, Mono Lake Rests, 1965: A Summary of Theoretical Prediction Methods and Some Comparisons with

Experimental   Data.   National   Technical   Information
Service, Springfield, Va.
Black,   L.   G.   1958.   Report   on   the   Mono   Lake
Investigation.   Prepared for the Los Angeles Department
of Water and Power. 74 pp.
Mono Lake Committee.   Various issues.   The Mono Lake
Newsletter.
Schwan, T. G., J. J. Oprandy, and A. J. Main.   In
preparation.   Ecological and laboratory studies of Mono
Lake virus.   In preparation for Journal of Wildlife
Diseases.
Twain, M., and M. R. Hill.   1965.   Islands of Mono Lake.
California   Division   of   Mines   and   Geology   Mineral
Information Service 18(9):173-180.
U.S. Department of the Interior.   1973.   Geothermal
Leasing Program, Vol. 1. Promulgation of Leasing and
Operating   Regulations   (Final   Environmental   Impact
Statement).   National   Technical   Information   Service,
Springfield, Va.

Index

259